Drafting, Interpretation, and Amendment of Patent Claims

ONLINE-KURS ZUM BUCH

Als Nutzer*in dieses Buches haben Sie kostenlos Zugriff auf einen Online-Kurs, der das Buch optimal ergänzt und für Sie wertvolle digitale Materialien bereithält. Zugang zu diesem Online-Kurs auf einer Springer Nature-eigenen eLearning-Plattform erhalten Sie über einen Link im Buch. Dieser Kurs-Link befindet sich innerhalb der ersten Kapitel. Sollte der Link fehlen oder nicht funktionieren, senden Sie uns bitte eine E-Mail mit dem Betreff „Book+Course“ und dem Buchtitel an customerservice@springernature.com.

Online-Kurse bieten Ihnen viele Vorteile!

- Sie lernen online jederzeit und überall
- Mit interaktiven Materialien wie Quizzen oder Aufgaben überprüfen Sie kontinuierlich Ihren Lernfortschritt
- Die Videoeinheiten sind einprägsam und kurzgehalten
- Tipps & Tricks helfen Ihnen bei der praktischen Umsetzung der Lerninhalte
- Ihr Zertifikat erhalten Sie optional nach erfolgreichem Abschluss

Thomas Heinz Meitinger

Drafting, Interpretation, and Amendment of Patent Claims

Procedures before the German Patent and Trade Mark Office, the European Patent Office, and the Federal Patent Court

Springer

Thomas Heinz Meitinger
Meitinger Patentanwalts GmbH
Munich, Germany

ISBN 978-3-662-73621-0 ISBN 978-3-662-73622-7 (eBook)
https://doi.org/10.1007/978-3-662-73622-7

Translation from the German language edition: "Formulierung, Auslegung und Änderung von Patentansprüchen" by Thomas Heinz Meitinger, © Der/die Herausgeber bzw. der/die Autor(en), exklusiv lizenziert an Springer-Verlag GmbH, DE, ein Teil von Springer Nature 2025. Published by Springer Berlin Heidelberg. All Rights Reserved.

This book is a translation of the original German edition "Formulierung, Auslegung und Änderung von Patentansprüchen" by Thomas Heinz Meitinger, published by Springer-Verlag GmbH, DE in 2025. The translation was done with the help of an artificial intelligence machine translation tool. A subsequent human revision was done primarily in terms of content, so that the book will read stylistically differently from a conventional translation. Springer Nature works continuously to further the development of tools for the production of books and on the related technologies to support the authors.

This Springer imprint is published by the registered company Springer-Verlag GmbH, DE, part of Springer Nature.
The registered company address is: Heidelberger Platz 3, 14197 Berlin, Germany

Preface

The patent claims define the scope of protection of a patent. Therefore, the patent claims represent the most important part of a patent. The wording of the patent claims deserves the utmost attention, as inadequate formulation can significantly reduce the potential scope of protection. Furthermore, errors in claim drafting can hardly be corrected later on.

It is important to understand how claims are to be interpreted. Patent claims are construed from the perspective of the person skilled in the art under patent law. The person skilled in the art is a hypothetical individual who, using various interpretative approaches, determines the meaning of a claim. In particular, the skilled person interprets the meaning of each individual term in the claim and places it in the context of the entire application documents.

After submission to the patent office, the possibility of amending patent claims is limited. It is particularly important to ensure that there is no "inadmissible extension" or "broadening of the scope of protection."

This technical book describes all phases of a technical intellectual property right, from claim drafting prior to filing with the patent office, to the interpretation of claims by patent offices and courts, and the possibilities for amending an already filed claim during opposition, appeal, and nullity proceedings.

Munich
in August 2025

Dr. Thomas Heinz Meitinger

Competing Interests The author has no conflicts of interest relevant to the content of this manuscript.

Contents

About the Author

Dr. Thomas Heinz Meitinger is a German and European patent attorney. He is the managing director of Meitinger Patentanwalts GmbH. Meitinger Patentanwalts GmbH is a medium-sized patent law firm based in Munich. After studying electrical engineering in Karlsruhe, he initially worked as a development engineer. Subsequent positions included roles as production manager and technical director in medium-sized companies. Dr. Meitinger regularly publishes scientific articles, writes specialist books on intellectual property law, and gives lectures on patent, trademark, and design law. Dr. Meitinger holds the degrees Dipl.-Ing. (Univ.) and Dipl.-Wirtsch.-Ing. (FH). He also holds the following master's degrees: LL.M., LL.M., MBA, MBA, M.A., and M.Sc.

List of Abbreviations

ABl. EPA	Official Journal of the European Patent Office
BeckOK	Beck's Online Commentaries
BeckRS	Beck case law (decision database from Beck-Online)
BGH	Federal Court of Justice (Bundesgerichtshof)
BGHZ	Decisions of the Federal Court of Justice in Civil Matters
BlPMZ	Journal for Patent, Design, and Trademark Matters
BPatG	Federal Patent Court (Bundespatentgericht)
DPMA	German Patent and Trademark Office (Deutsches Patent- und Markenamt)
EPA/EPO	European Patent Office (Europäisches Patentamt)
EPA G	Case law of the Enlarged Board of Appeal of the European Patent Office
EPA T	Case law of the Technical Boards of Appeal of the European Patent Office
EPC	European Patent Convention (also EPÜ: Europäisches Patentübereinkommen)
GRUR	Gewerblicher Rechtsschutz und Urheberrecht
GRUR-RS	GRUR Collection of Case Law
GRUR Int	GRUR International
InstGE	Decisions of the lower courts on intellectual property law
Mitt.	Communications of the German Patent Attorneys (Mitteilungen der deutschen Patentanwälte)
NJOZ	New Legal Online Journal
NJW	New Legal Weekly
PatG	(German) Patent Act
RG	Imperial Court (Reichsgericht)
RGZ	Collection of Decisions of the Reichsgericht
Rn.	Margin number
UPC	Unified Patent Court (also EPG: Einheitliches Patentgericht)
VPP-Rdbr	VPP Newsletter

List of Figures

List of Tables

Introduction 1

How the online course enriches the book "Drafting, Interpretation, and Amendment of Patent Claims":
As a reader of this book, you can access the associated online course free of charge. The course complements the content of this book and also provides guidance for successful implementation in everyday practice.

To use it, please follow these instructions:

Use the following link: sn.pub/tipdd1 to access the accompanying online course. If you are using it for the first time, please register on the learning platform.

In this first chapter of the book, the essential aspects of patent law are explained, enabling the reader to apply the instructions for action described here in the correct context.

1.1 Claims

The patent claims are the central element of a patent. The patent claims define the scope of protection, i.e., the extent of the exclusive right granted to the patent holder.[1] The remainder of the patent may only serve to support and clarify the understanding of the scope of protection.[2] Furthermore, the wording of the claims forms the basis for the patent office's search for prior art.

Claims describe in a single sentence what the patent holder claims as the exclusive scope of use. From the wording of the claim, one can directly derive the embodiments

[1] Section 14, sentence 1 of the Patent Act.

[2] Section 14, sentence 2 of the Patent Act.

T. H. Meitinger, *Drafting, Interpretation, and Amendment of Patent Claims*,
https://doi.org/10.1007/978-3-662-73622-7_1

for which the patent holder has an exclusive right of prohibition against any unauthorized party. An authorized party would be, for example, a licensee who has purchased the right to use the invention by paying a license fee. In addition, the right of prohibition cannot be asserted against persons who use the technical teaching in a private context.[3]

1.2 Description

The description comprises the entire application documents except for the claims. The description may in particular include drawings that depict different embodiments. In one section of the description, the drawings are explained in detail, with each depicted element and its interaction with the other elements being described. In addition, the description may include the prior art and describe its disadvantages. The disadvantages of the prior art lead directly to the objective of the invention.

1.3 Person Skilled in the Art

The fictional figure of the person skilled in the art in patent law serves to interpret the claims. Everything that the average skilled person would understand from the wording of a claim is considered to be encompassed by that claim. Therefore, if an embodiment is disclosed from the perspective of the average skilled person using their technical knowledge and expertise, the patent law criterion of sufficiency of disclosure is met, even if it may appear incomprehensible to a layperson.

1.4 Problem-Solution Approach

The problem-solution approach was developed by the European Patent Office and serves to systematically and transparently assess the inventive step of patent claims.[4] In particular, its application is intended to exclude a distorting ex post analysis.

The problem-solution approach is carried out in three steps. In the first step, the document is identified that provides the best starting point for arriving at the invention to be examined. This closest prior art should belong to the same technical field as the claimed invention and share as many features as possible with the invention. In this case, the

[3] Section 11, no. 1 of the Patent Act.

[4] EPO, https://www.epo.org/de/legal/guidelines-pct/2025/g_vii_5.html, accessed on 22.6.2025.

Table 1.1 Problem-Solution Approach

Step 1	Closest Prior Art	The document is identified that provides the best starting point for arriving at the invention. This document in particular shares the most features with the invention
Step 2	Objective Technical Problem	Based on the closest prior art, the problem to be solved in order to arrive at the invention is determined
Step 3	Examine whether the invention is obvious based on the closest prior art with the help of the objective technical problem	Could the skilled person have arrived at the invention from the document of the closest prior art if the objective technical problem had indicated which effects needed to be achieved? (Could-would test: And would the skilled person have combined the technical teachings of the necessary documents?)

fewest structural and functional modifications are required compared to the invention under examination.[5]

In the second step, the problem to be solved by the skilled person in order to achieve the technical teaching of the invention is determined. For this purpose, the distinguishing features are identified, i.e., those by which the closest prior art differs from the invention. The technical effect produced by these distinguishing features is determined. The skilled person must therefore modify the technical teaching of the closest prior art so that the technical effect of the distinguishing features is achieved.[6]

The objective technical problem does not have to coincide with the problem set by the inventor. It may therefore be necessary to amend the formulation of the technical problem in the application documents. A new formulation of the technical problem during the grant procedure is always permissible, as the applicant cannot be prevented from adapting the objective technical problem if a new legal situation arises as a result of the prior art identified by the patent office.[7]

In the third step, it must be determined whether there is a combination of prior art that discloses the claimed invention. It must also be examined whether such a combination would have been used by a skilled person. For this, the could-would test is to be applied. If both can be confirmed, the inventive step of the invention under examination must be denied. Table 1.1 shows the problem-solution approach.

[5] EPO, https://www.epo.org/de/legal/guidelines-pct/2025/g_vii_5_1.html, accessed on 22.6.2025.
[6] EPO, https://www.epo.org/de/legal/guidelines-pct/2025/g_vii_5_2.html, accessed on 22.6.2025.
[7] EPO, https://www.epo.org/de/legal/guidelines-pct/2025/g_vii_5_2.html, accessed on 22.6.2025.

1.5 Could-Would Test

The Could-Would Test requires that only those combinations of prior art documents be considered that “fit together”. A skilled person would not combine technical teachings that contradict each other. Therefore, if the inventor has combined technical teachings that, from the perspective of a skilled person at the time, would have been mutually exclusive, an invention resulting from such a combination is based on inventive step.[8]

The Could-Would Test thus simulates the approach of a skilled person. It is assumed that a skilled person works in a goal-oriented manner and does not combine documents at random. Therefore, it is not only relevant that a combination of prior art documents could be made (“could”), but also whether the skilled person would actually make this particular combination (“would”).[9]

1.6 Patent Grant Procedure

The official examination procedure serves to crystallize out allowable patent claims. To this end, the patent examiner will first identify the relevant prior art. In a second step, the examiner will in particular assess the novelty and inventive step of the independent claims in light of the prior art found.

Figure 1.1 shows the chronological sequence of the grant procedure, illustrating the special situation in which a request for examination is filed simultaneously with the submission of the application documents. In this case, the patent office aims to send the applicant the first official communication regarding the patentability of the application within the one-year priority period.[10] The applicant is then able, based on the patent office’s assessment of patentability, to decide whether it is worthwhile to seek patent protection for the invention in foreign countries. The applicant is entitled to a right of priority within the first year after filing the application documents with a patent office. Within this year, the applicant can file national or regional patent applications anywhere in the world, with the filing date of the first application being recognized as the priority date.[11]

[8] EPO, https://www.epo.org/de/legal/guidelines-epc/2024/g_vii_5_3.html, accessed on 22.6.2025.

[9] EPO T 2783; EPO T 257/98; EPO T 35/04; EPO T 623/97; EPO T 558/00.

[10] Right of priority: Article 4 Paris Convention for the Protection of Industrial Property (Paris Convention, Pariser Verbandsübereinkunft zum Schutz des gewerblichen Eigentums, PVÜ); one-year priority period for subsequent foreign patent applications: Article 4 C (1) Paris Convention.

[11] Article 4 Section C Paragraph 1 Paris Convention for the Protection of Industrial Property (Paris Convention, PVÜ).

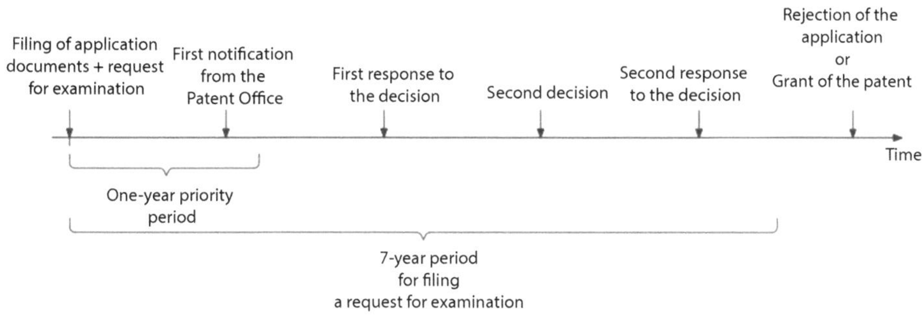

Fig. 1.1 Chronological sequence of the grant procedure

In principle, the request for examination at the German Patent Office can be filed within a period of seven years.[12] However, the German Patent and Trademark Office will interpret a late filing of the request for examination as an indication that the applicant is not interested in a prompt conclusion. The office will therefore give priority to applications for which the request for examination was filed earlier.

There is no official examination procedure under the Utility Model Act. Only a request for search can be filed, so that the prior art relevant for assessing legal validity is determined.[13] The request can be filed by the utility model holder or any third party.

1.7 Opposition Procedure

Through an opposition procedure, any person may file an opposition against the grant of a patent. There is no need to demonstrate a legal interest. The opposition procedure is conducted before the respective patent office and is usually based on lack of novelty and inventive step of the independent claims. An opposition can only be filed within nine months after publication of the grant of the patent.

Table 1.2 lists the possible grounds for opposition on which a valid opposition may be based.

[12] Section 44 Paragraph 2 Sentence 1 Patent Act.

[13] Section 7 Paragraph 1 Utility Model Act.

Table 1.2 Grounds for Opposition

Ground for Opposition	Meaning	Legal Basis
Lack of novelty	The subject matter of the claim was made available to the public before the filing or priority date	Section 21 Paragraph 1 No. 1 in conjunction with Section 3 Paragraph 1 Sentence 1 Patent Act or Article 100(a) in conjunction with Article 54 Paragraph 1 EPC
Lack of inventive step	The invention is obvious to the skilled person in light of the prior art	Section 21 Paragraph 1 No. 1 in conjunction with Section 4 Sentence 1 Patent Act or Article 100(a) in conjunction with Article 56 Sentence 1 EPC
Insufficient disclosure	The skilled person is not able, based on the disclosure of the patent specification, to produce or carry out the subject matter of the claim	Section 21 Paragraph 1 No. 2 Patent Act or Article 100(b) EPC
Impermissible extension	The technical teaching of the patent exceeds the disclosure of the originally filed application documents	Section 21 Paragraph 1 No. 4 Patent Act or Article 100(c) EPC
Unauthorized removal	The invention was not filed for patenting by the inventor or their legal successor	Section 21 Paragraph 1 No. 3 Patent Act

1.8 Nullity Proceedings

The nullity proceedings are judicial proceedings before the Federal Patent Court by which a patent or the German part of a European patent can be challenged by means of an action.[14] A nullity action can be filed by any person. A nullity action is not admissible during the opposition period of the contested patent and as long as opposition proceedings are pending.[15]

The grounds for nullity correspond to those of the opposition proceedings, with the addition of the ground of extension of the scope of protection.

Table 1.3 lists the possible grounds for revocation on which an admissible nullity action can be based.

[14] Section 81 paragraph 1 sentence 1 Patent Act.

[15] Section 81 paragraph 2 sentence 1 Patent Act.

Table 1.3 Grounds for Revocation

Ground for Revocation	Meaning	Legal Basis
Lack of novelty	The subject matter of the claim was made available to the public before the filing or priority date	Section 22 paragraph 1 in conjunction with Section 21 paragraph 1 no. 1 in conjunction with Section 3 paragraph 1 sentence 1 Patent Act
Lack of inventive step	The invention is obvious to the skilled person in view of the prior art	Section 22 paragraph 1 in conjunction with Section 21 paragraph 1 no. 1 in conjunction with Section 4 sentence 1 Patent Act
Insufficient disclosure	The skilled person is not able, on the basis of the disclosure in the patent specification, to produce or carry out the subject matter of the claim	Section 22 paragraph 1 in conjunction with Section 21 paragraph 1 no. 2 Patent Act
Unallowable extension	The technical teaching of the patent exceeds the disclosure of the originally filed application documents	Section 22 paragraph 1 in conjunction with Section 21 paragraph 1 no. 4 Patent Act
Extension of scope of protection	The scope of protection of the patent was impermissibly extended during the opposition proceedings	Section 22 paragraph 1 Patent Act
Unauthorized removal	The invention was not filed for patenting by the inventor or his legal successor	Section 22 paragraph 1 in conjunction with Section 21 paragraph 1 no. 3 Patent Act

1.9 Patent Infringement Proceedings

By filing a lawsuit before a specialized patent litigation chamber of a regional court, a patent holder can assert their claims against a patent infringer. In particular, the patent holder may base their lawsuit on a claim for injunctive relief to prevent future unlawful acts and on a claim for damages to financially compensate for past patent infringements. As a rule, an attempt is made to enforce the patent holder's rights out of court through a warning procedure before filing a lawsuit.

Table 1.4 lists the claims that a patent holder can assert in patent infringement proceedings.

Three different types of patent infringement can be distinguished. In the case of literal infringement, the infringing product fulfills all features as formulated in the claim. Equivalent patent infringement does not fulfill the features of the claim exactly and is

Table 1.4 Claims of the Patent Holder

Patent Holder's Claim	Meaning	Legal Basis
Claim for Injunctive Relief	The patent holder can assert a claim for injunctive relief against future patent infringements in cases of risk of recurrence and in cases of a substantiated risk of first-time infringement against highly probable imminent patent infringements.	Section 139 Paragraph 1 Sentences 1 and 2 Patent Act
Claim for Damages	The patent infringer must compensate the patent holder for the damage caused.	Section 139 Paragraph 2 Sentence 1 Patent Act
Claim for Information	The infringer must provide information about the extent of the damage to enable the patent holder to calculate damages and to allow the patent holder to take action against further patent infringers.	Section 140b Paragraph 1 Patent Act
Claim for Destruction	The patent holder is entitled to the destruction of infringing products and equipment that serve solely to manufacture infringing products.	Section 140a Paragraphs 1 and 2 Patent Act
Claim for Recall	The patent holder is entitled to have infringing products permanently removed from the distribution chain.	Section 140a Paragraph 3 Patent Act
Claim for Inspection	The patent holder is entitled to inspect a third party's premises if there is no other way to prove the patent infringement.	Section 140c Paragraph 1 Sentence 1 Patent Act
Publication of the Judgment	In special cases, the patent holder is entitled to the publication of the infringement court's judgment.	Section 140e Sentence 1 Patent Act

therefore only recognized under strict conditions. In indirect infringement, only part of the features of the claim are fulfilled by the patent infringement.

Literal Patent Infringement

In literal or verbatim patent infringement, every feature of the claim, as described, is realized by the infringing product. The infringing product thus represents exactly the subject matter of the claim, without any technical modification.

Equivalent Patent Infringement

Equivalent patent infringement exists if a claim is not realized verbatim, but the substituted features are characterized by an equivalent and equally effective nature compared to the original features.

The substituted features that replace the original features must meet the following criteria to constitute equivalent patent infringement:

- Equivalent effect: the substituted features must act in the same way.
- Equivalence: From the perspective of a skilled person, the substituted features and the original features must appear equivalent.
- Obviousness: Replacing the original features with the substituted features must require at most routine skill. If the modification constitutes an inventive step, equivalent patent infringement is excluded.

Indirect Patent Infringement

Indirect patent infringement occurs when the infringer supplies essential means of the subject matter of the patent to a third party, who uses these means to realize the patented product. A prerequisite for indirect patent infringement is that the infringer knows or it is obvious to them that these means are intended for patent infringement.[16] However, indirect patent infringement only exists if the means are not products that are generally available on the market. Nevertheless, the following applies: If the means are generally available products and the infringer supplies them to the third party so that a patent infringement is intentionally committed, a patent infringement still exists.[17]

[16] Section 10 Paragraph 1 Patent Act.

[17] Section 10 Paragraph 2 Patent Act.

Claim Drafting

2

The drafting of claims prior to submitting the application documents to the patent office represents an initial attempt to describe a grantable scope of protection. It is advisable to choose the broadest possible claim wording. The initial formulation of the set of claims should include independent claims that, in light of the known prior art, are just barely novel. The assessment of inventive step by the patent office is difficult to predict. It is recommended to await the examiner's reasoning in the official communication and, if necessary, to restrict the claims in order to achieve patentability.

2.1 Principles of Claim Drafting

A claim should have the quality of a recipe, specifying the features required to produce the invention, or containing the steps necessary to carry out the inventive method.

A product claim has a structure in which the subject matter is first designated, or a characterization of the item to be protected is provided with "Device for ". This is followed by a listing of the individual features that the inventive device comprises. Care should be taken to ensure that a logical connection is established between the individual features.

For a manufacturing or application process, for example, the following can be formulated:

"Method for producing an electric vehicle comprising the steps:
manufacturing a body,
installing an electric motor in the body,
installing a battery to supply the electric motor with electrical energy, …".

T. H. Meitinger, *Drafting, Interpretation, and Amendment of Patent Claims*,
https://doi.org/10.1007/978-3-662-73622-7_2

The claims must be executable by a person skilled in the art. A claim must therefore be drafted so clearly and completely that the skilled person can reproduce the invention.[1] Otherwise, the respective claim is not allowable due to lack of executability.

2.2 Claims of the Patent Application

The claims of a patent application represent an initial attempt at formulation. It is advisable to try to describe, with the claims, a scope of protection that covers subject matter likely to be patentable. This does not mean that only very specific and detailed subject matter should be claimed, which would probably lead to a patent quickly. In this case, one would likely have claimed too narrow a scope of protection and would later have to contend with competitors who could circumvent the patent with a minor modification of one's own invention.

It is rather advantageous for the initial claim formulation to already describe the fundamental subject matter around which the technical teaching of the invention revolves. In this case, one avoids the patent office conducting multiple searches for prior art during the grant procedure, as the fundamental patent request has changed. If the patent office examiner has to search for prior art multiple times, this leads to significant delays and a rapid grant of the patent will not be possible.

The initial claim formulation should be as broad as possible. If the claims of the application are just barely novel in light of the known prior art, the claims have been strategically drafted correctly. In this case, one can be confident that the entire scope of protection not belonging to the prior art is being claimed.

The applicant must be aware that their aim must be to obtain the broadest possible scope of protection. Otherwise, they will very quickly be confronted with imitators. A patent that can be circumvented with little technical effort is even worse than having no patent at all, as it discloses the technical solution to competitors without preventing imitation.

2.3 Claims of the Patent

The claims of the patent determine the scope of protection of the intellectual property right.[2] The patent claims therefore represent the most important part of a patent. The description and drawings serve only to aid understanding and proper interpretation of the claims.[3] The description and drawings can in no case establish their own scope of protection.

[1] Section 34, paragraph 4 of the Patent Act

[2] Section 14, sentence 1 of the Patent Act.

[3] Section 14, sentence 2 of the Patent Act.

Table 2.1 Claim Categories

Claim Category	Variants of the Claim Category
Device	Device for manufacturing
	Device as object (e.g., machine)
Process	Process for manufacturing
	Process for use

2.4 Types of Claims and Claim Categories

There are independent and dependent claims. The first claim in a set of claims is the independent "main claim," which contains the essential features of the invention.[4] The other independent claims are referred to as parallel claims, which also contain essential features of the invention.[5] The independent claims may relate to different subject matters, but must be linked by a single common inventive concept.[6]

The independent claims describe a distinct subject matter of the invention, for example, a machine, a device for manufacturing the machine, a process for manufacturing the machine, or a process for using the machine. Parallel claims do not have to refer back to preceding claims. However, a parallel claim may indeed include the features of one or more preceding claims by using a formulation such as "wherein the process is carried out with a device according to any one of claims 1 to 7" or "wherein the device is manufactured/applied by a process according to claims 1 to 7."

Table 2.1 provides an overview of the different claim categories. Essentially, devices (products) and processes can be distinguished. A set of claims may contain both device and process claims, with device claims relating to the object or the thing itself, such as a machine or a building. In addition, process claims may be included, which describe the manufacture or use of the object in process steps. A claim for a "process for use" can be a use claim for a substance or product. Until now (Federal Patent Court, decision of 28.10.2002, 5 W(pat) 25/01, Mitt. 2004, 266), a use was regarded as a process claim. In a subsequent decision by the Federal Court of Justice (BGH decision of 5 October 2005, X ZB 7/03—Utility Model for Medicinal Products), the use of a substance was classified as a product claim, since in a use claim only an already existing property of a product is utilized. The classification of a use claim as a device or process claim is therefore disputed.

Table 2.2 shows the relationship between the terms independent/dependent claim and main, parallel, and sub-claim. The main claim and the parallel claims describe a distinct subject matter or process and each claim its own scope of protection. The dependent

[4] Section 9, paragraph 4 of the Patent Ordinance.

[5] Section 9, paragraph 5 of the Patent Ordinance.

[6] Section 34, paragraph 5 of the Patent Act.

Table 2.2 Types of Claims

Type of Claim	
Independent claim	Main claim
	Parallel claim/parallel independent claim
Dependent claim	Sub-claim

claims refer to the independent claims and specify their subject matter. Dependent claims do not establish their own scope of protection. A dependent claim always refers back to an independent claim and belongs to the claim category of the referenced independent claim.

Figure 2.1 provides an example of a set of claims to illustrate the distinction between types of claims and claim categories. In particular, it is shown that a referenced sub-claim must always belong to the same patent category as the respective main or parallel claim.

The disadvantage of a process claim is that proving infringement can be difficult. For example, if a patented manufacturing process is used in a country where the patent is not in force to produce goods for supply to the protected domestic market, a patent infringement occurs. However, there is no right of inspection in the country without patent protection, and therefore proof of infringement will not succeed.

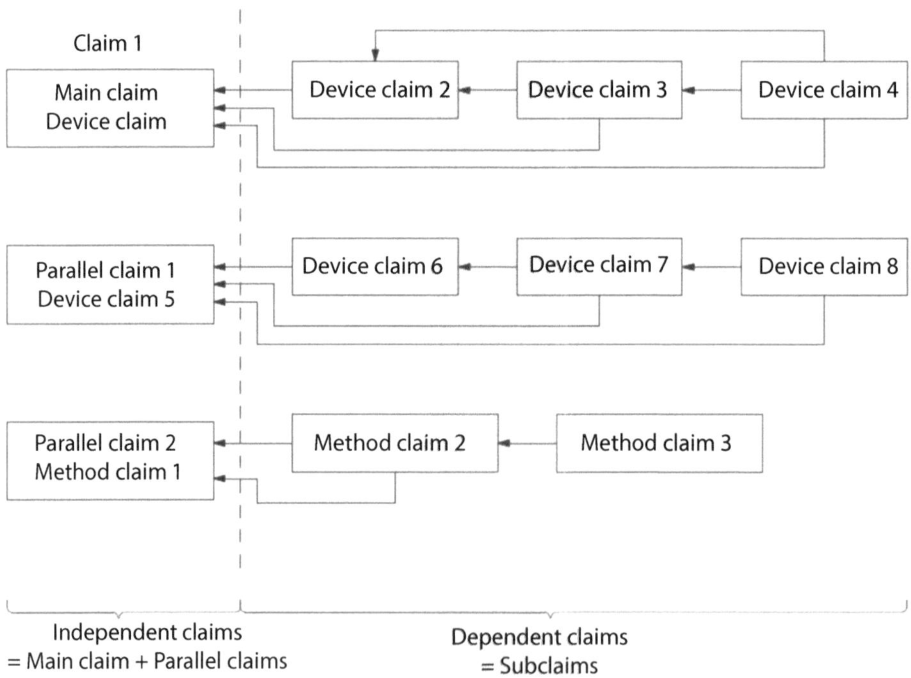

Fig. 2.1 Types of Claims and Claim Categories

However, according to Section 9, sentence 2, no. 3 of the Patent Act, the scope of protection of a patented process also extends to the products manufactured directly by the process. It must be demonstrated that these are products manufactured by the process. It must therefore be shown that the products possess a particular property that they have acquired solely as a result of the patented process. The products must have acquired characteristic properties through the process.[7] If the products can only be manufactured by this process, the proof is already provided. The extension of the scope of protection of the patented process to the directly manufactured products is also effective if the products themselves are not patentable.[8]

The required directness of the products is given if no further processing or treatment steps have been carried out.[9] According to Section 9, sentence 2, no. 3 of the Patent Act, the products must therefore be end products. This restriction is intended to prevent an excessive extension of the scope of protection to a sequence of products.

Section 9, sentence 2, no. 3 of the Patent Act was specifically created to close the gap in protection that arises when a manufacturing process protected domestically is relocated to a country without patent protection for the purpose of supplying the domestic market.[10]

2.5 Drafting Independent Claims

Figure 2.2 shows the basic structure of an independent claim, illustrated here in the two-part format. In the two-part format, the features are separated into those known from the prior art and those features responsible for the inventive effect, whereby features known from the prior art may also be used for this purpose. The first group of features is referred to as the "preamble" and the second group as the "characterizing portion." The distinction can sometimes be difficult, as, for example, all features may be equally known and of equivalent quality. In such cases, the one-part claim format should be chosen and no separation made within the claim.

When drafting an independent claim as an apparatus claim, the features that are absolutely necessary to produce the inventive subject matter should be described. For a method claim, the individual steps should be listed in the correct order to describe the manufacture or use. Care should be taken not to describe the advantages of the invention, but rather the individual elements that are actually required to reproduce the invention.

[7] Higher Regional Court Düsseldorf, judgment of 10.4.2005—U (Kart) 44/01; Regional Court Düsseldorf InstGE 7, 70 Video Signal Coding I.

[8] Swiss Federal Court GRUR Int 96, 1059, 1060.

[9] Higher Regional Court Karlsruhe InstGE 11, 15 SMD resistor; Higher Regional Court Düsseldorf NJOZ 2010, 1781 interframe dropping.

[10] Federal Court of Justice GRUR 09, 1142, MP3 player import.

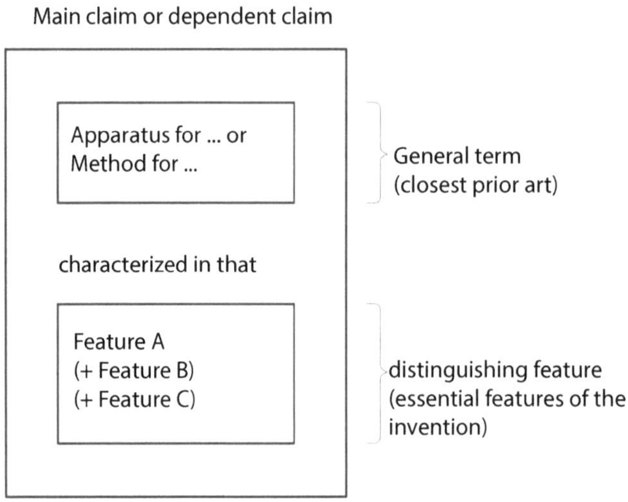

Fig. 2.2 Main claim or auxiliary claim

The independent claims should then be checked to ensure that they contain only those features that are absolutely necessary. Any features that are not strictly required must be removed from the independent claims. Only the necessary features should be included in the independent claims so that the scope of protection defined by the independent claims is as broad as possible. The "superfluous" features removed from the independent claims can be used to draft dependent claims.

The patent office regularly requires the applicant to draft the claims in a two-part format. The two-part claim format is characterized by a "preamble" and a "characterizing portion," with both parts separated in the claim by "characterized by" or "characterized in that". The preamble should contain the features known from the prior art, and the characterizing portion lists the features that represent the inventive development over the known prior art.[11] Preferably, the preamble describes those features that are contained in the closest prior art.[12] Since the document considered closest is usually the one that shares the most features with the invention, it can be assumed that the features in the preamble are known from the prior art and the features in the characterizing portion represent the essential inventive features.

Once the independent claims have been drafted, a comparison with the prior art identified in the search must be carried out. If the subject matter of a claim can be derived

[11] Section 9(2) sentence 1 of the Patent Ordinance or Rule 43(1) EPC.

[12] DPMA, https://www.dpma.de/docs/formulare/patent/p2796.pdf, accessed on 11.7.2025; BeckOK PatR/Cimniak, 31st ed., 15.1.2024, EPC Article 84 para. 29.

from a document in the prior art, the claim is not novel. In this case, a feature not disclosed in the relevant document must be added to establish novelty.

The question of inventive step for a claim is difficult to answer. It is therefore not possible to reliably predict how the patent office will assess a claim. It is advisable to wait for the patent office's opinion on inventive step in the examination report.

2.6 Drafting Dependent Claims

Dependent claims may include additional features that serve various purposes. This allows further specific embodiments of the invention to be described. A dependent claim may alternatively describe a specific numerical value of a parameter or a numerical range within which the invention works particularly well.

2.7 Structure of a Set of Claims

The structure of a set of claims is essentially determined by the independent claims to which the dependent claims refer back.

Figure 2.3 provides an example of the structure of a set of claims, distinguishing between independent and dependent claims. The independent claims determine the scope

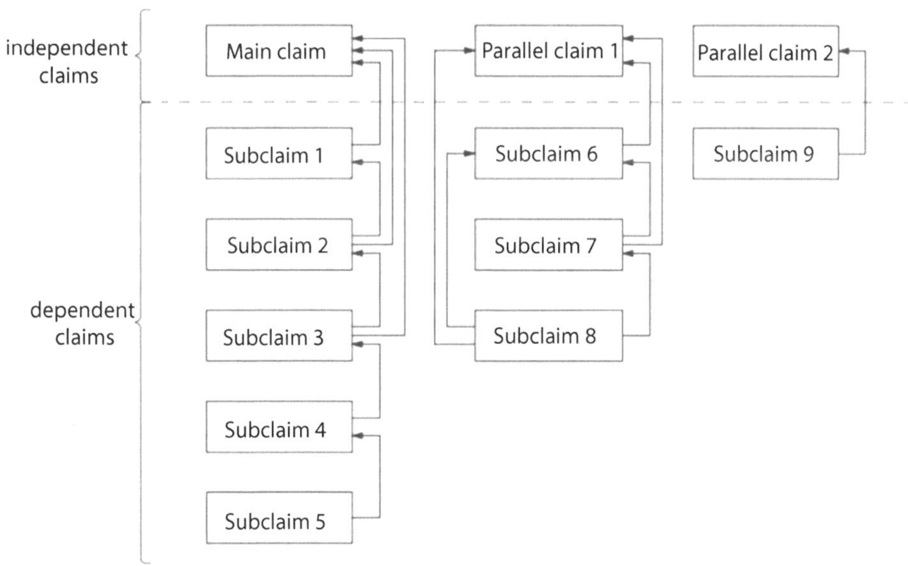

Fig. 2.3 Structure of a set of claims

of protection of the patent, with each independent claim defining a separate scope of protection. The dependent claims contain all the features of the respective referenced independent claim and additional features, thereby describing specific embodiments of the subject matter of the independent claim. The dependent claims therefore describe a subrange of the scope of protection of the referenced independent claim.

The independent claims often belong to different patent categories. In addition, they must meet the requirement of unity. Unity is present if the independent claims relate to a common inventive concept.[13] The independent claims must therefore have the essential features of the invention in common.

Auxiliary/parallel claims may refer to preceding claims.[14] However, auxiliary claims do not have to refer to preceding claims, in contrast to dependent claims, and auxiliary claims define their own scope of protection. This also does not apply to dependent claims, which only further describe a part of the scope of protection of an independent claim.

2.8 Subjects of a Set of Claims

When formulating a set of claims, one should also consider which subjects arise from the dependencies. Sets of claims often contain claims that are cascaded and refer back to each other:

1. Device for… (**Claim 1**—main claim/independent claim)
2. Device according to claim 1, wherein… (**Claim 2**—dependent claim)
3. Device according to claim 1 or 2, wherein… (**Claim 3**—dependent claim)
4. Device according to any one of the preceding claims… (**Claim 4**—dependent claim)
5. Device according to any one of the preceding claims… (**Claim 5**—dependent claim)
6. Device according to any one of the preceding claims… (**Claim 6**—dependent claim)

From the respective dependency of a claim on all preceding claims, it follows in this example that the following subjects are claimed:

From **Claim 1** results 1 subject:
Subject 1: features of claim 1

From **Claim 2** results 1 subject:
Subject 2: features of claim 2 and features of claim 1

[13] Section 34(5) Patent Act.
[14] Section 9(5) sentence 3 of the Patent Ordinance.

From **Claim 3** result 2 subjects:
Subject 3: features of claim 3 and features of claim 1
Subject 4: features of claim 3 and features of claim 2 and features of claim 1

From **Claim 4** result 4 subjects:
Subject 5: features of claim 4 and features of claim 1
Subject 6: features of claim 4 and features of claim 2 and features of claim 1
Subject 7: features of claim 4 and features of claim 3 and features of claim 1
Subject 8: features of claim 4 and features of claim 3 and features of claim 2 and features of claim 1

From **Claim 5** result 8 subjects:
Subject 9: features of claim 5 and features of claim 1
Subject 10: features of claim 5 and features of claim 2 and features of claim 1
Subject 11: features of claim 5 and features of claim 3 and features of claim 1
Subject 12: features of claim 5 and features of claim 3 and features of claim 2 and features of claim 1
Subject 13: features of claim 5 and features of claim 4 and features of claim 1
Subject 14: features of claim 5 and features of claim 4 and features of claim 2 and features of claim 1
Subject 15: features of claim 5 and features of claim 4 and features of claim 3 and features of claim 1
Subject 16: features of claim 5 and features of claim 4 and features of claim 3 and features of claim 2 and features of claim 1

From **Claim 6** result 16 subjects:
Subject 17: features of claim 6 and features of claim 1
Subject 18: features of claim 6 and features of claim 2 and features of claim 1
Subject 19: features of claim 6 and features of claim 3 and features of claim 1
Subject 20: features of claim 6 and features of claim 3 and features of claim 2 and features of claim 1
Subject 21: features of claim 6 and features of claim 4 and features of claim 1
Subject 22: features of claim 6 and features of claim 4 and features of claim 2 and features of claim 1
Subject 23: features of claim 6 and features of claim 4 and features of claim 3 and features of claim 1
Subject 24: features of claim 6 and features of claim 4 and features of claim 3 and features of claim 2 and features of claim 1
Subject 25: features of claim 6 and features of claim 5 and features of claim 1
Subject 26: features of claim 6 and features of claim 5 and features of claim 2 and features of claim 1

Subject 27: features of claim 6 and features of claim 5 and features of claim 3 and features of claim 1
Subject 28: features of claim 6 and features of claim 5 and features of claim 3 and features of claim 2 and features of claim 1
Subject 29: features of claim 6 and features of claim 5 and features of claim 4 and features of claim 1.
Subject 30: features of claim 6 and features of claim 5 and features of claim 4 and features of claim 2 and features of claim 1
Subject 31: features of claim 6 and features of claim 5 and features of claim 4 and features of claim 3 and features of claim 1
Subject 32: features of claim 6 and features of claim 5 and features of claim 4 and features of claim 3 and features of claim 2 and features of claim 1

The number of subjects described by a set of claims, when each claim refers back to all preceding claims, is: $2^{\text{claim number}-1}$. For a set of 10 claims, this results in 512 different subjects.

2.9 Scopes of Protection of a Set of Claims

Figure 2.4 shows the visual representation of the respective scopes of protection resulting from the following set of claims:

1. Device for… (Claim 1—independent claim)
2. Device according to claim 1, wherein… (Claim 2—dependent claim)
3. Device according to claim 1 or 2, wherein… (Claim 3—dependent claim)
4. Device according to any of the preceding claims… (Claim 4—dependent claim)

The independent claim (claim 1) encompasses all devices within the outermost frame of Fig. 2.4. Claim 2 must fall within the scope of protection of claim 1, since claim 2 is only dependent on claim 1. At most, the scope of protection of claim 2 could correspond to that of the independent claim if, for example, the additional features of claim 2 merely represent a different wording of the features of the independent claim or are non-technical features. Claim 3 is dependent on claims 1 and 2, so claim 3 covers a scope of protection that it shares only with the independent claim, and a scope that it shares with claim 2, and thus together with claim 2 and the independent claim. Claim 4 is dependent on claims 1, 2, and 3, so claim 4 has a scope of protection that it shares only with the independent claim, a scope it shares with the independent claim and claim 2, a scope it shares with the independent claim and claim 3, and a scope it shares with the independent claim and claims 2 and 3. The scope of protection of claim 4 can therefore be divided into four distinct sub-areas.

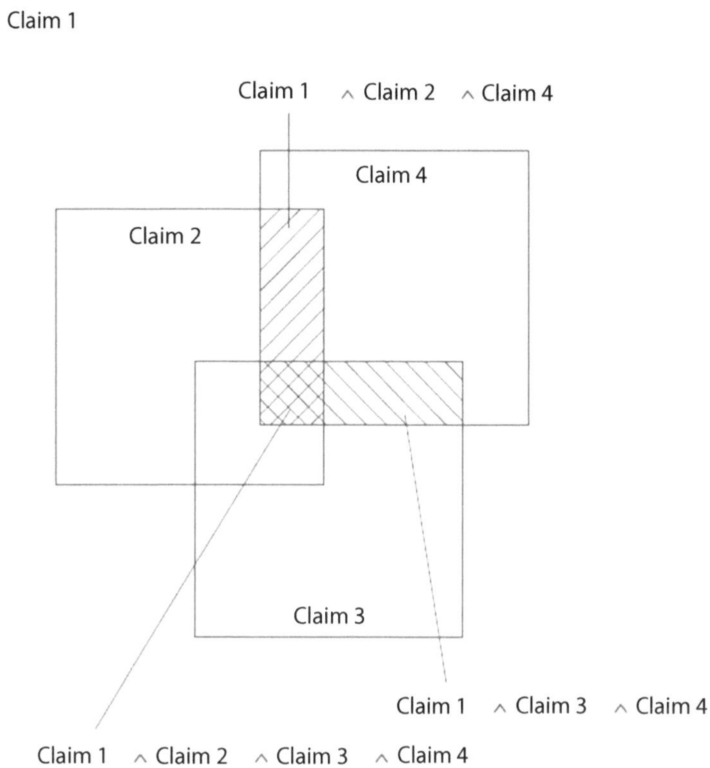

Fig. 2.4 Scopes of protection of a first set of claims

It is entirely possible for a dependent claim not to refer back to all preceding claims, but for example only to the immediate predecessor. This may be necessary in particular if a term is used in the relevant claim that is introduced only in the immediately preceding claim. Claim 4 of the following set of claims is such a dependent claim, referring only to the preceding claim:

1. Device for…
2. Device according to claim 1, wherein…
3. Device according to claim 1 or 2, wherein…
4. Device according to claim 3…

Figure 2.5 illustrates the respective scope of protection described by the individual claims of the set.

Independent claim 1 encompasses any subject matter within the outer frame. The scope of protection of claim 2 lies entirely within that of the independent claim and

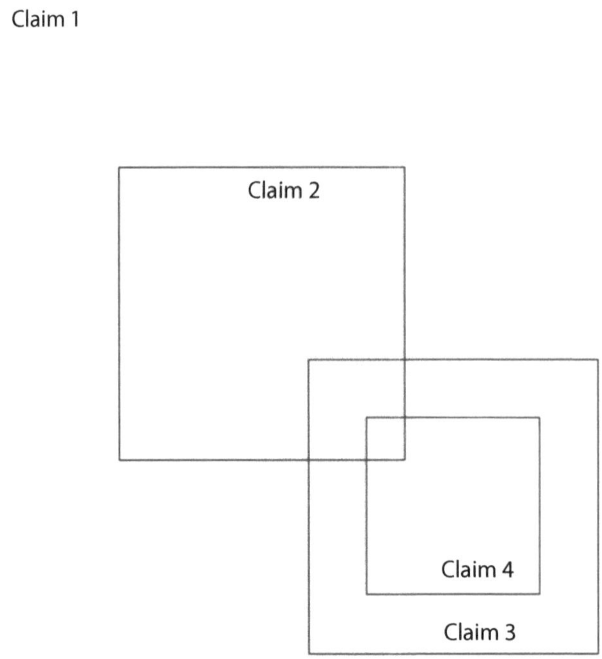

Fig. 2.5 Scopes of protection of a second set of claims

selects a section, further defining the subject matter of this section by additional features. Claim 3 overlaps the scope of protection of claim 2 and lies within the scope of claim 1. Claim 4 is only dependent on claim 3 and thus carves out a section of the scope of claim 3, further specifying the subject matter of this section by additional features. The scope of protection of claim 4 therefore falls entirely within that of claim 3.

Another alternative set of claims is given if the dependent claims refer only to one claim, in particular the independent claim. The following set of claims represents this variant:

1. Device for… (Claim 1—independent claim)
2. Device according to claim 1, wherein… (Claim 2—dependent claim)
3. Device according to claim 1, wherein… (Claim 3—dependent claim)
4. Device according to claim 1, wherein… (Claim 4—dependent claim)

In this case, combinations of the features of the dependent claims with each other are excluded, so that the scope of protection of the claims is as shown in Fig. 2.6. In this set of claims, for example, a subject matter that includes both the features of claim 2 and those of claim 3 is not covered.

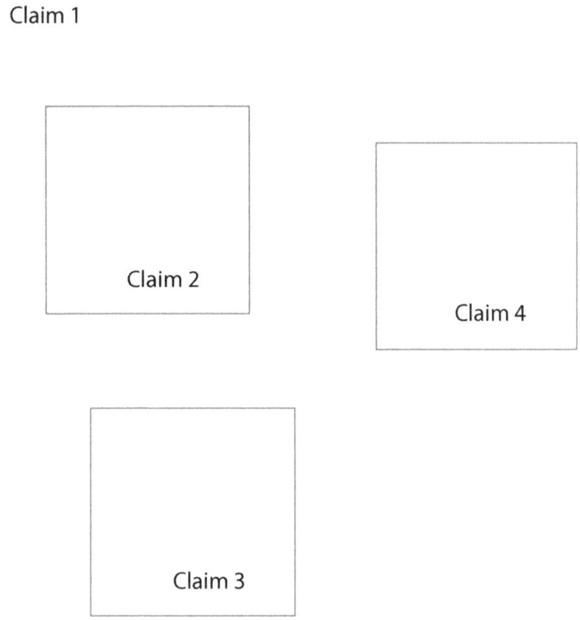

Fig. 2.6 Scopes of protection of a third set of claims

Figure 2.6 shows the set of claims with three dependent claims, each referring only to the independent claim. There are therefore no overlaps between the scopes of protection of the dependent claims. In this set of claims, the independent claim serves as the bracket ensuring the unity of the claims. If, for example, the independent claim must be deleted in opposition or nullity proceedings, three non-unitary scopes of protection result, which are no longer linked by a single common inventive concept.[15]

Figure 2.7 shows a representation of a set of claims, making it clear that the dependent claims constitute a specification of the scope of protection of the main claim or of the respective referenced subclaim. The subclaims therefore describe particular embodiments of the scope of protection.

The dependent claims can be particularly significant in cases of cross-border use of the technical teaching of the patent. For example, if features 1, 2, and 3 are implemented in a country where the patent is not in force, there is no patent infringement with respect to the main claim. However, if the additional feature 5 of subclaim 2 is not also implemented in the patent-free country, but rather in the country where the patent is protected, the use of the patent is at least completed in the protected country, resulting in a patent infringement.

[15] Section 34 paragraph 5 Patent Act.

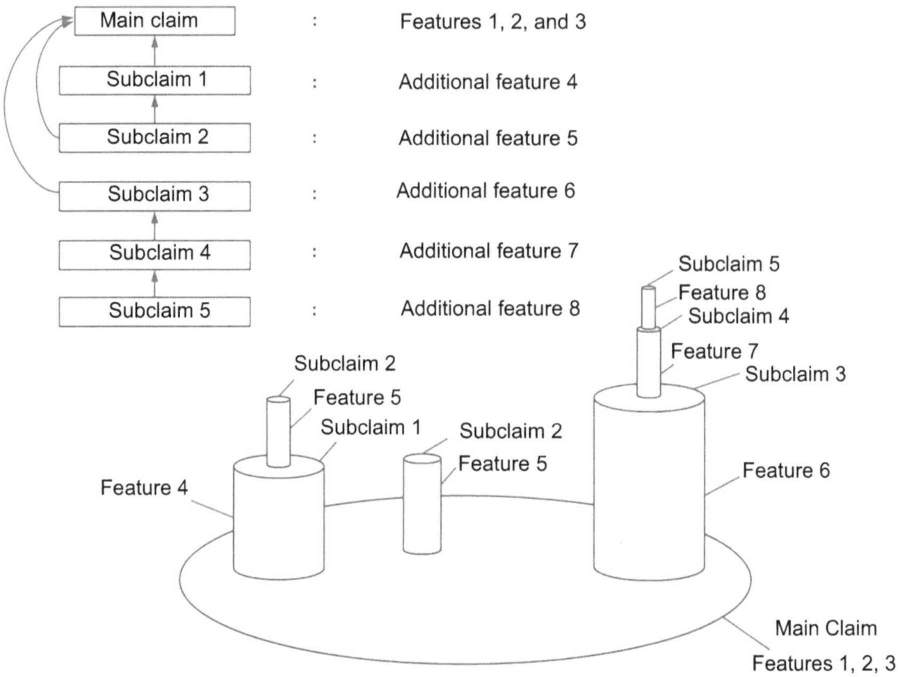

Fig. 2.7 Visualization of a set of claims

2.10 Clarity

Clarity of the terms used in the claims, as well as clarity of the overall statement of the claims, is a prerequisite for legal certainty. Any third party must be able to determine from the wording of the claims what is permitted and what is covered by the exclusive right of the patent. Clarity is required from the perspective of the skilled person. A claim is clear if its meaning is "immediately and unambiguously" apparent to the skilled person.[16] If a claim formulation is clear, there is no need to replace the respective terms with even more precise formulations during the grant, opposition, or nullity proceedings. Unnecessary replacements should be consciously avoided in order to eliminate the risk of inadmissible extension.

A court cannot refuse to interpret patent claims due to lack of clarity. On the contrary, in cases of ambiguous claim formulations, it is precisely the task of the competent court to clarify the appropriate scope of protection. This can already be inferred from the fact that the interpretation of a patent claim is the clarification of the legal question regarding

[16] EPO G 02/88 OJ EPO 90, 93—friction-reducing additive/Mobil Oil.

its scope of protection. Legal questions raised must be clarified by the court seized of the matter.[17]

Claims should be formulated as generally as possible in order to cover the greatest possible number of embodiments. However, the claim formulation must be clear. Otherwise, the claim becomes ambiguous and can no longer serve as the basis for an injunction.

Lack of clarity is, alongside lack of novelty or inventive step, the most frequent deficiency asserted by the patent office. However, lack of clarity of the claims is not expressly required in the German Patent Act.[18] It is only stipulated that the invention must be disclosed clearly in the application.[19] Furthermore, the Patent Act provides that the patent claims must specify what protection is claimed.[20] From this, it can only be concluded that clarity is also required in claim formulation; otherwise, there would be no sufficient description of the invention.

In opposition or nullity proceedings, lack of clarity cannot be asserted as a ground for revocation.[21] This also applies to the European opposition procedure,[22] even though the European Patent Convention requires a clear claim formulation.[23] A review of granted claims for clarity is therefore not permissible in opposition or nullity proceedings.[24]

2.11 Breadth of the Claim

The applicant has the right to claim the entire scope of protection by means of an appropriate claim formulation, to the extent permitted in view of the prior art.[25] Thus, any number of embodiments can be encompassed by the scope of protection. A very broad scope of protection is not objectionable for that reason alone.[26]

[17] German Federal Court of Justice (BGH) March 31, 2009—X ZR 95/05—Road construction machine.

[18] Federal Patent Court (BPatG), judgment of June 24, 2015–15 W (pat) 9/13—Polyurethane foam; BPatG 20 W (pat) 71/04 of April 15, 2009; BPatG 15 W (pat) 33/08 of December 16, 2013—Battery monitoring device; BPatG; 20 W (pat) 8/14 of April 7, 2014—Electric control device; BPatG 11 W (pat) 32/13 of December 15, 2014 – Cooking device.

[19] Section 34 paragraph 4 Patent Act.

[20] Section 34 paragraph 3 number 3 Patent Act.

[21] BGH, judgment of December 15, 2020—X ZR 180/18—Disc brake; BGH, judgment of October 27, 2015—X ZR 11/13, GRUR 2016, 361 para. 31—Joint tape; judgment of January 19, 2016—X ZR 141/13, GRUR 2016, 475 para. 39—Receptor tyrosine kinase I; EPO, decision of March 24, 2015—G 3/14, OJ 2015, A102 paras. 73 ff.

[22] EPO, https://www.epo.org/de/legal/guidelines-epc/2024/d_v_5.html, accessed on 14.8.2025; Article 100 EPC.

[23] Article 84 sentence 2 EPC.

[24] BGH March 18, 2025—X ZR 30/23.

[25] EPO G 01/98 OJ EPO 2000, 111—transgenic plant/Novartis II.

[26] BGH GRUR 04, 47—bubble-free rubber sheet I.

However, a very broad claim formulation carries the risk that a large number of prior art documents will be identified by the patent office during the grant procedure and may stand in the way of patent grant. A large number of prior art documents is disadvantageous because the applicant's own set of claims must be distinguished from all these documents and, for every possible and, from the skilled person's perspective, reasonable combination of prior art, must be based on an inventive step. Furthermore, a very broad claim formulation entails the risk that only after several official communications and responses by the patent proprietor will an allowable version of the set of claims be found.

A very narrow formulation can advantageously lead to a rapid grant of the patent. However, this may result in a scope of protection that is too small, inviting imitators to implement a workaround by making minor modifications to the technical teaching of the patent. In addition, the public is informed about the applicant's technical know-how without, in return, obtaining a sufficiently broad scope of protection.

A proven rule, therefore, is to formulate the claims of a patent application such that they are at least novel over the prior art identified by the applicant's own search. Since inventive step is always a matter of perspective or argumentation, this criterion should, except in clear situations, initially be disregarded. With this approach, the entire scope of protection not covered by the prior art is fully claimed. In addition, a claim formulation is already chosen that is likely to be relatively close to an allowable claim.

2.12 Precision and Completeness

A claim describes the "recipe" for implementing the claimed invention. Consequently, the claim must include all features or process steps necessary to produce or apply the invention. Therefore, a claim must be "complete".

The individual features must also be drafted in such a way that a person skilled in the art can carry them out.[27] If the claim cannot be carried out by a person skilled in the art, the patent may be revoked in opposition proceedings[28] or declared invalid in nullity proceedings.[29]

[27] Section 34, paragraph 4 of the Patent Act.

[28] Section 61 in conjunction with Section 21, paragraph 1, no. 2 of the Patent Act.

[29] Section 81 in conjunction with Section 22, paragraph 1, in conjunction with Section 21, paragraph 1, no. 2 of the Patent Act.

2.13 Enforceability and Verifiability

As a rule, the patent holder bears the burden of proof that their patent is being infringed. Therefore, when drafting their claims, the patent holder should ensure that they can generally meet this burden of proof or fulfill it in a straightforward manner.

Enforceability is particularly ensured if proof of patent infringement can be provided by simple means or without the need to inspect the premises of the potential infringer. For this reason, product claims relating to a commercially available product are preferable to patent-protected manufacturing process claims. Determining that a patented process is being used in a competitor's operation is often difficult. It is therefore advisable always to include a product claim in the set of claims.

If the claim is a manufacturing claim, it may be necessary to enforce a right of inspection to determine that the infringing act is taking place at the unauthorized party's premises.[30] If production takes place in a country where the patent is not in force, the patent holder will not be granted a right of inspection.

The issue of manufacturing in a country where the patent is not in force for the purpose of supplying the protected domestic market is recognized by the legislator. Therefore, patent protection has been extended to products directly obtained by a process.[31] Such products do not themselves have to meet the patentability criteria of novelty and inventive step.[32] In addition, the legislator provides for a reversal of the burden of proof to protect the patent holder from having to prove infringement of a manufacturing claim in a country where the patent is not in force. However, the reversal of the burden of proof only applies to products that have not yet been offered on the market.[33]

2.14 Connection Between the Features of a Claim

The features of a claim must be in a logical or physical relationship. It is not permissible for a claim to include objects that are not structurally or logically related, thereby claiming several independent objects in a single claim.

[30] Section 140c, paragraph 1, sentence 1 of the Patent Act.

[31] Section 9, sentence 2, no. 3 of the Patent Act.

[32] Schulte/Rinken, PatG, 10th edition, Section 9, margin no. 96.

[33] Section 139, paragraph 3, sentence 1 of the Patent Act.

If an independent claim contains unrelated features, the non-essential features should be removed from the claim, as they evidently do not need to interact with the remaining features to achieve the inventive effect. Advantageously, this can result in a broader scope of protection.

2.15 Support in the Description

The claims must be "supported" by the description.[34] Such "support in the description" is required by both German and European patent law.[35] This "requirement for a description" is intended to ensure that the applicant was in genuine possession of the invention. It also ensures that the invention can be carried out. The description must therefore relate to the claims and must not appear unrelated to them or even contradict them.

In particular, for and/or combinations of a first feature and a second feature in a claim, the description should include embodiments with the first or the second feature, as well as with a combination of the first and second features.

The description of the claims should especially be used to preemptively counter the objection of lack of enablement by providing concrete explanations of the subject matter of the claims.[36] However, it should be borne in mind that the description must be taken into account when interpreting the claims.[37] Therefore, if a term is described or defined in such a way that it gives the term in a claim a narrower meaning, the patent holder must be held to that definition, even if the technical language attributes a broader meaning to the term.

2.16 Essential Features

The first claim, the main claim, must describe the essential features that are absolutely necessary to implement the invention.[38] Therefore, the main claim should not include detailed information about specialized embodiments. Overly detailed manufacturing instructions and particularly specific embodiments and construction guidelines can

[34] Article 84, sentence 2 EPC.

[35] Article 84, sentence 2 EPC. In the German Patent Act, Section 34, paragraph 4 requires that the invention be described in the application.

[36] Section 34, paragraph 4 of the Patent Act or Article 83 EPC.

[37] Section 14, sentence 2 of the Patent Act or Article 69, paragraph 1, sentence 2 EPC; EPO G 1/24, 18.6.2025, https://www.epo.org/en/boards-of-appeal/decisions/g240001ex1, accessed on 21.8.2025.

[38] Section 9, paragraph 4 of the Patent Ordinance.

instead be disclosed in the description.[39] However, this does not apply to features that are indispensable for carrying out the invention. Such features are considered essential and must be described in the main claim.[40]

2.17 Functional Features

Functional features describe an invention not by its structural, physical properties, but by the intended effect or function to be fulfilled. Claims consisting of functional features are referred to as "means-plus-function" claims for devices or "steps-plus-function" claims for methods. Functional features provide an abstraction that leads to embodiments being described that include all physical features that fulfill the same function. When using functional features, there is a risk that the patent office will question enablement. Therefore, concrete examples should be provided in the description to address this objection.

Functional features are regularly used to describe the subject matter or method of a claim by using the phrase "device for ..." or "method for ...".

Examples of the use of functional features:

"1. Device for transporting goods and people, comprising:
a motor for driving the device."

In this claim, the phrase "device for transporting goods and people" particularly encompasses automobiles, motorcycles, ships, airplanes, rockets, and trains.

2.18 Descriptive Features

Purely descriptive features that describe features already present in the relevant claim cannot provide any additional technical contribution. For better readability of the claim, such duplications should be deleted. Such features, which provide additional explanations, can be included in the description.[41]

Optional features are often integrated into the claim using terms such as "in particular", "preferably", "advantageously", "for example", or "where appropriate". They do not contribute to the scope of protection but merely describe a subset thereof.

---

[39] Federal Court of Justice GRUR 72, 707, 13—Streckwalze I.

[40] EPO T 0805/93 OJ EPO 98 Special Edition 36; EPO T 0133/85 OJ EPO 88, 441; EPO T 0622/90 OJ EPO 92 Special Edition 70.

[41] Werner, Communications of the German Patent Attorneys, 78, 226; Oettinger, Communications of the German Patent Attorneys, 79, 51; BGH BIPMZ 54, 24—Multiple Clamp.

2.19 Alternative Features

A claim may contain alternative features that are connected by an "or" conjunction. It must be ensured that the alternatives agree in their essential features so that they are characterized by a common inventive concept and the criterion of unity is maintained.

2.20 Features from a Different Category

It is permissible to include method steps in apparatus claims or to include tangible or structural features in method claims.[42] However, this is only permissible if otherwise it would not be possible to describe the subject matter of the claim. Furthermore, it must remain obvious which category the claim belongs to.

Examples of claims containing features from a different category are, in particular, "product-by-process" claims and purpose-limited substance claims for pharmaceuticals (second medical indication according to Section 3(4) Patent Act or Article 54(5) EPC).

2.21 Irrelevant Features

The independent claims should be checked before filing the application documents to determine whether they contain irrelevant or overly restrictive features. Irrelevant features that are not absolutely necessary for the implementation of the invention should be moved to a dependent claim. The same applies to unnecessarily restrictive features that do not concern the essence of the invention.

2.22 Non-Technical Features

Non-technical features do not belong in a patent claim. A rare exception exists if non-technical features are required, together with technical features, to produce the inventive effect.

A combination of technical and non-technical features can arise, in particular, in software inventions.[43] It is permissible for non-technical features to be included in claims in order to describe the technical features. However, when examining novelty and inventive step, the non-technical features are not taken into account. Therefore, only those features

[42] BGH GRUR 86, 163—boron-containing steels.

[43] Features that describe "software as such" are considered non-technical: Section 1(3) No. 3 in conjunction with (4) Patent Act.

that contribute to the technical character of the invention can contribute to patentability.[44] Features that serve to solve a non-technical problem, for example because aesthetic requirements are met, are to be excluded from consideration for patentability.[45]

Non-technical terms should therefore be avoided in claims. Examples of non-technical terms are: "aesthetic", "theoretical", "mathematical", "business activity", "business model", or "marketing measure". In most cases, non-technical features arise from the exclusion criteria of the Patent Act.[46]

2.23 Prior Art Search

Before drafting claims, a search for relevant prior art should always be conducted. The purpose of such a search is, in particular, to avoid preparing a patent application that is anticipated by a prior art document. In such a case, the entire effort for the patent application would be in vain.

Furthermore, a two-part claim formulation can only be drafted in light of the prior art. A two-part claim comprises a preamble, the features of which are known from the prior art, and a characterizing portion, the features of which essentially embody the invention.

If relevant documents are found during a prior art search, these should be studied thoroughly. The documents can be seen as an opportunity to further develop one's own technology. In this way, one's own technology can be even better adapted to customer requirements, potentially resulting in a superior advanced technology that gives the applicant a significant lead over competitors.

In particular, engaging with the prior art may reveal special embodiments of one's own technology, which can lead to valuable dependent claims.

A prior art search is therefore recommended to prepare for the examination procedure and to draft the patent application in such a way, based on the prior art, that a clear distinction from the prior art is achieved. Additionally, a search offers the opportunity to further develop one's own technology and thus make it more successful.

2.24 Development of a Proprietary Technology Based on a Search

It is entirely possible to encounter a situation where one is aware that there is a need for technical advancement in a particular technical field. If it is a technical field that has not experienced any significant progress for some time, for example, the transfer of

---

[44] EPO, https://www.epo.org/de/legal/guidelines-epc/2024/g_vii_5_4.html, accessed on 23.6.2025.

[45] EPO T 641/00, EPO G 1/19 (Comvik approach).

[46] Section 1(3) Patent Act.

advanced technologies from other technical fields could lead to increased productivity and customer benefits.

The starting point for developing a proprietary technology can be a search for the relevant documents in the respective technical field. Each document should be thoroughly analyzed in this process.

The development of a proprietary technology can, in particular, proceed in five steps.

Step 1: What Problems Need to be Solved?

It is important to understand which problems in the state of the art need to be solved. The specific problems of a technical field must be identified. For example, a typical problem in a given technical field might be sealing a combustion chamber or avoiding friction losses that lead to heat generation. In such cases, the applicant should therefore provide a good sealing solution or a technically sophisticated solution for heat dissipation.

Step 2: How Does the State of the Art Work?

The mode of operation must be studied in detail to identify the advantages and disadvantages. By analyzing the documents, the different solution methods become apparent, allowing the most advantageous one to be selected for one's own purposes. It may be possible to combine individual features or steps from the technical teachings of different documents to arrive at a proprietary superior technology.

Step 3: What Objective Does the State of the Art Address?

In this step, it becomes clear what the actual objective is that needs to be solved in order to produce a marketable product. For market success, it may be necessary not only to solve the typical problems of the specific technical field but also to create additional customer value. This step is therefore important to determine which objectives the market wants to see solved. This can help avoid solving problems for which there is no market demand.

The objective of the state of the art may, in particular, be to create additional benefits to increase customer satisfaction.

Step 4: What are the Disadvantages of the State of the Art?

Identifying the disadvantages of the state of the art provides another starting point for developing a proprietary technology. Simply recognizing and avoiding or mitigating the disadvantages of the state of the art can lead to a proprietary technology.

Step 5: What Alternative Solutions are Available?

At the latest from this step onward, the development of a proprietary technology begins. Ideally, the inventor at this point possesses the extensive technical knowledge acquired through the analysis of the state of the art. Thus, they have become an expert in this technical field and are therefore well prepared to create an advanced new technology.

This sequence of five steps demonstrates that intensive engagement with the state of the art provides the best conditions for developing a superior proprietary technology. In

Table 2.3 Functions of Searching the State of the Art

State of the Art	Assessment of Patentability	Technology Development
Function 1	Identify the closest state of the art to draft claims in two-part form	Develop proprietary technology
Function 2	Assessment of novelty	Improve proprietary technology
Function 3	Assessment of inventive step	Identify special embodiments

Table 2.4 Sequence of Steps for Technology Development

Development of a Technology Based on the State of the Art		
Step 1	What problems does the state of the art solve?	The typical problems arising in the specific technical field are identified.
Step 2	How does the state of the art work?	Which mechanisms of action are used in the state of the art?
Step 3	What objective does the state of the art address?	What customer benefit does the state of the art aim to provide—greater comfort, ease of handling, or cost-effective production?
Step 4	What are the disadvantages of the state of the art?	What does the state of the art fail to achieve adequately?
Step 5	What alternative solutions are available?	Are there alternative technical teachings that are less disadvantageous? Can know-how transfer from other technical fields succeed in achieving a novel solution?

particular, opportunities may arise to refine an existing technology or at least to develop further specific embodiments of one's own technology, which can, for example, successfully serve market niches.

Table 2.3 presents the various functions of searching the state of the art, with three functions each for the patent law assessment of one's own invention and for the development, improvement, and expansion of the technical teaching of the invention.

Table 2.4 summarizes the sequence of steps for developing a proprietary technology.

2.25 Closest Prior Art

The closest prior art is the document from the state of the art that represents the most promising starting point for reproducing the invention. The closest prior art should originate from the same technical field as the invention and pursue at least a similar objective.[47]

[47] EPO, https://www.epo.org/de/legal/guidelines-epc/2024/g_vii_5_1.html, accessed on 10.7.2025.

The European Patent Office regularly considers the document that shares the most features with the invention as the closest prior art. The closest prior art is therefore a document that requires the fewest structural and functional modifications to arrive at the invention.[48]

2.26 Assessment of Novelty

A claim is only allowable if it meets the absolute novelty criterion of patent law. "Absolute" means that any document must be considered that was made available to the public anywhere in the world, at any time before the filing date of the patent application describing the invention in question. It does not matter in which language it is written or where it was published. It is sufficient if a single skilled person understood the invention and was not bound by a confidentiality agreement, for the invention to be considered known to the public and therefore no longer novel.

If a document was made available to the public before the filing date or the priority date of the patent application, it is considered relevant prior art, and a claim of the patent application or the patent is only patentable if the subject matter of the claim cannot be derived from the document.[49]

2.27 Assessment of Inventive Step

A claim must not only be novel. In addition, the technical teaching must exhibit a minimum distance from the prior art. A teaching that can be obtained by the mere routine skills of an average skilled person should not justify a monopoly right through a patent.

The problem-solution approach, which originates from the European Patent Office and is also applied in a slightly modified form at the German Patent and Trademark Office, describes procedural steps to determine the inventive step of a claim in a verifiable manner.[50] The problem-solution approach consists of three steps, which are carried out in sequence to make a well-founded decision on the inventive step of the technical teaching under examination.

Problem-Solution Approach

The problem-solution approach comprises three steps, beginning with the identification of the closest prior art. Next, it is determined what distinguishes the technical teaching of the closest prior art from the invention under examination. In a third step, it is considered whether the skilled person could arrive at the invention under examination by

---

[48] EPO T 606/89.

[49] EPO, https://www.epo.org/de/legal/guidelines-epc/2024/g_vi_1.html, accessed on 22.7.2025.

[50] EPO, https://www.epo.org/de/legal/guidelines-pct/2025/g_vii_5.html, accessed on 22.7.2025.

Table 2.5 Problem-Solution Approach

Problem-Solution Approach		
Step 1	Identify the closest prior art	The closest prior art is that which originates from the same technical field as the invention, pursues at least a similar objective, and shares the most features with the invention.
Step 2	Which features of the invention are not present in the closest prior art, and what technical effect do these distinguishing features produce?	The closest prior art does not exhibit all features of the invention; otherwise, the invention would not be novel. The task of the skilled person is therefore to arrive at these distinguishing features, and he will do so by searching for prior art documents in which the technical effect produced by these distinguishing features is described.
Step 3	Is there a combination of documents that the skilled person would consider, and whose combination discloses the invention (Could-Would Test)?	In this step, it must be assessed which documents the skilled person would combine, as he would regard the relevant technical teachings as combinable. If such a meaningful combination of prior art documents results in the invention, it is not inventive.

combining one or more further documents from the prior art. Only those documents are considered that the skilled person would regard as combinable with the closest prior art (Could-Would Test).

Step 1: First, the closest prior art must be identified. The closest prior art is characterized by originating from the same technical field as the invention under examination and sharing the most features with the invention.

Step 2: The features of the invention that are not described by the closest prior art are determined. In addition, the technical effect resulting from the distinguishing features is established. The skilled person would therefore have the objective task, starting from the closest prior art, of finding prior art documents that produce this technical effect.

Step 3: If the skilled person can find prior art documents that he would combine with the closest prior art and that achieve the technical effect of the distinguishing features, the invention is not inventive. If he cannot, the invention is based on an inventive step.

Table 2.5 shows the successive steps of the problem-solution approach for examining an invention for inventive step.

2.28 Product-by-Process Claim

A product-by-process claim should not be confused with a product under Section 9 sentence 2 no. 3 of the Patent Act. A product under Section 9 sentence 2 no. 3 of the Patent Act is a product that results directly from a patented process, which itself does not have to be patentable. In contrast, a product-by-process claim must be novel and inventive in order to establish an exclusive right.

For a product under Section 9 sentence 2 no. 3 of the Patent Act, the special properties of the product serve to provide evidence that the patented process has been infringed. The product itself does not have to be patentable. In contrast, in a product-by-process claim, the steps of the manufacturing process must result in properties of the product that make it independently patentable.[51]

A product-by-process claim should generally be avoided. The formulation of a product claim as a product-by-process claim is only permissible if no other claim formulation is possible.[52] It is also advisable for the applicant to describe a claim with its tangible features wherever possible, rather than with process steps of manufacture, since proving that process steps have been applied by a patent infringer is often difficult.

In a product-by-process claim, the processing by means of the process must therefore impart properties to the product that produce the particular technical effect and establish the patentability of the product.[53]

If a product-by-process claim is present, the corresponding manufacturing process is generally also patentable and should be included in the set of claims as an independent claim.

2.29 Selection Invention

A selection invention claims a subset from a larger numerical range. This subset may represent numerical values for mixing ratios, dimensions, or temperatures.

In selection inventions, it is often found that range limits are set for economic rather than technical reasons. From the perspective of the patent office, this constitutes an arbitrary determination that does not lead to a legally valid formulation. Instead, the range limits must result from an inventive insight in order to be accepted as an invention by the patent office.

The inventor develops their invention, which functions at certain parameter values. To deter competition, ranges are then defined around the identified parameter values and described in a claim. It is advisable not to define ranges that are too broad, so as not to claim embodiments that cannot be attributed to the invention. If the ranges are very large, the patent office may get the impression that the applicant is seeking an unjustified advantage. It is sensible to set such limits within which the inventive effect is still reliably present, at least in a diminished form. Numerical values justified solely by economic considerations should be avoided.

[51] EPO, https://www.epo.org/de/legal/guidelines-epc/2024/f_iv_4_12.html, accessed on 1.7.2025, BGH, Judgment of 17 July 2025—X ZR 40/23—Spreizdübel II.

[52] EPO, https://www.epo.org/de/legal/guidelines-epc/2024/f_iv_4_12.html, accessed on 1.7.2025.

[53] BGH, Judgment of 17 July 2025—X ZR 40/23—Spreizdübel II.

A selection invention is considered novel if the selected range is a small part of a large range in the prior art. In addition, the selected range must be sufficiently distant from the ranges known in the prior art.[54] If the numerical values of the subrange result in an inventive and unexpected effect, inventive step can be assumed.

2.30 Formal Requirements for Patent Claims

The "Ordinance on Proceedings in Patent Matters before the German Patent and Trademark Office" (Patent Ordinance) contains the formal requirements for patent claims in § 9.

In principle, the applicant may choose whether to draft their patent claims in the two-part or one-part format. In the two-part format, the features are included in a preamble and a characterizing portion.[55] The preamble describes the features that are derived from the prior art. The characterizing portion contains the features which, together with those of the preamble, produce the desired inventive effect.[56] The characterizing portion therefore contains the features that are significant for producing the inventive effect. The preamble and the characterizing portion are to be separated by "characterized in that" or "characterized by".[57]

The individual claims of a set of claims are to be numbered consecutively with Arabic numerals.[58] The claims must not contain textual references to the description or the drawings. References such as "see drawing 2", "as explained in description page 2", or "as shown in drawing 1" are not permitted.[59] On the other hand, reference signs referring to individual elements of the drawings are to be included in the claims.[60] The reference signs are to be placed in parentheses.

2.31 Number of Claims

In German proceedings, the first 10 claims are included in the filing fee. From the 11th claim onwards, a claim fee of €20 must be paid for each additional claim.[61]

In European proceedings, the first 15 claims are included, and only from the 16th claim onwards is a separate claim fee due for each additional claim. For the 16th to the

[54] EPO, https://www.epo.org/de/legal/guidelines-epc/2025/g_vi_7.html, accessed on 1.7.2025.
[55] § 9 paragraph 1 sentence 1 Patent Ordinance.
[56] § 9 paragraph 2 sentence 1 Patent Ordinance.
[57] § 9 paragraph 2 sentence 2 Patent Ordinance.
[58] § 9 paragraph 7 Patent Ordinance.
[59] § 9 paragraph 8 Patent Ordinance.
[60] § 9 paragraph 9 Patent Ordinance.
[61] Annex to § 2 paragraph 1 fee item no. 311 050 Patent Costs Act, as of November 2025.

50th claim, a fee of €275 is payable for each claim. From the 51st claim onwards, €685 must be paid for each additional claim (Article 2 paragraph 1 no. 15 Fee Schedule (Rule 45 paragraph 1, Rule 71 paragraph 4, Rule 162 paragraph 1), as of November 2025).

It is therefore advisable to formulate a maximum of 10 claims in German proceedings and a maximum of 15 claims in European proceedings. If there are additional special embodiments that could be described in a dependent claim, these can be included in the description. The applicant is free, during the grant procedure, to extract features from the description instead of from a dependent claim in order to distinguish a main claim from the prior art. The applicant therefore does not suffer any disadvantage by including special embodiments of their invention in the description rather than in dependent claims.

The maximum number of claims should be fully utilized, as the dependent claims are examined by the official examiner during the grant procedure, so that the applicant receives an official assessment of patentability for the subject matter of the dependent claims.

2.32 One-Part and Two-Part Claim Drafting

In the two-part claim format, the features known from the prior art are arranged in a first section of the claim, the preamble, and the features which, together with those of the preamble, produce the inventive effect are described in a characterizing portion. Both sections are separated by "characterized in that" or "characterized by".

Separation of the patent claim into a preamble and a characterizing portion is only possible if the invention is not a combination invention. In a combination invention, all features of the claim interact equally and equivalently, so that it is not possible to distinguish between features that are significant for the invention and those that are merely required. In addition, all features may be equally known from the prior art. In this case, it is the combination itself, and not individual features, that constitutes the invention.[62] In such cases, one should not forcibly attempt a two-part claim formulation, which would always be "incorrect", but instead draft the claim in one-part form.

2.33 References to the Description and Drawings

References in the claims to the description and the drawings are not permitted.[63] Such references are also superfluous, since according to § 14 sentence 2 Patent Act, for the interpretation of the claims, the description and the drawings must always be consulted,

[62] EPO, https://www.epo.org/de/legal/case-law/2022/clr_i_d_9_3_1.html, accessed on 11.7.2025.

[63] § 9 paragraph 8 Patent Ordinance.

even if the claim wording is clear and unambiguous in itself.[64] In addition, it is possible to refer to individual parts of drawings in the claims by means of reference signs. A literal reference, for example by "as can be seen in the drawing under …", can therefore be replaced by reference signs.

2.34 References to Other Documents

References in claims to external documents are not permitted, as they conflict with the public's need for legal certainty. The public must be able to determine the scope of protection with reasonable effort. Claims must therefore be clear in themselves, whereby recourse to the description and drawings of the relevant patent is reasonable (and required).

2.35 Reference Signs

Reference signs improve the comprehensibility of claims by referring to individual components or sections of the drawings of the patent application.[65] The use of reference signs does not limit the claimed scope of protection, but serves only for description.[66] The reference signs are to be placed in parentheses in the claims.

2.36 Numerical Values

Numerical values or parameters can serve to specify a feature more precisely. They therefore serve to clearly distinguish the invention from the prior art. Numerical values are particularly well suited for delimiting the invention from the prior art and should be included in a dependent claim.

However, numerical values are only accepted by the patent office as a means of distinguishing from the prior art if they possess an inventive character. The description should explain why the chosen numerical values represent a particular inventive contribution or why and what specific inventive effect is achieved by them.

[64] EPO G 1/24.

[65] § 9 paragraph 9 Patent Ordinance or Rule 43 paragraph 7 sentence 1 EPC.

[66] Rule 43 paragraph 7 sentence 2 EPC.

2.37 Qualifying Terms

A claim must be formulated clearly and unambiguously. This is particularly important to ensure that the claim can be carried out. Terms that qualify or relativize circumstances can jeopardize the feasibility of the respective claim, as they lead to a lack of precision.

It should be borne in mind that the skilled person will in any case take manufacturing tolerances into account, so that, from this perspective alone, additional qualification is usually unnecessary. Terms such as "approximately", "about", "roughly", "any", or "typically" should therefore be avoided when drafting claims.[67]

2.38 Indefinite Terms

Indefinite terms arise, for example, from word combinations such as "as small as possible", "as large as possible", "substantially round", or "substantially with a rough surface". Such word combinations do not allow for an unambiguous description of a circumstance. Moreover, they are often formulated as objectives and do not describe how something is manufactured or applied. Indefinite terms should not be used in claims.

The word "respectively" ("beziehungsweise" in German) is ambiguous, as it can mean "and", "or", "furthermore", "including", "in addition", "additionally", "with", "besides", and "depending on". Instead of "bzw.", a clear conjunction should be used.

2.39 Technical Terminology

Although a patent specification is its own lexicon, it is nevertheless preferable not to deviate from general usage or technical terminology. This helps to prevent linguistic confusion. Linguistic ambiguity can be exploited by an opposing party in contentious proceedings to assert lack of enablement as a defense or as grounds for revocation.

2.40 Contradictions

Contradictions in a patent application or patent specification pose a risk to successful patent grant, to defending the patent in opposition and nullity proceedings, and to enforceability in patent infringement proceedings. It is not a problem if the description contains more than is claimed as the scope of protection in the claims. It is also not a problem if the claim covers more embodiments than are included in the description. It is

[67] Article 84, sentence 2 EPC, Section 34(4) PatG.

sufficient if the application documents describe individual examples of the embodiments covered by the scope of protection.

However, contradictions between terms within an independent claim lead to ambiguity of the statement, which is always interpreted by a court to the detriment of the applicant. In the worst case, this can result in such a claim being denied any scope of protection in a patent infringement proceeding.

The claims should be "supported" by the description.[68] The description and the drawings are to be used for interpreting the claims.[69] If the description and the patent claims contradict each other, opportunities may arise for the opposing party of the patent proprietor to prevent the enforceability of the patent. For example, if features are described as essential to the invention in the description but are not included in the main claim, a contradiction already exists.

2.41 "Comprising" and "Having" versus "Consisting of"

The terms "comprising"/"having" and "consisting of" are interpreted differently by the Federal Patent Court and infringement courts. "Comprising" and "having" are interpreted as non-exhaustive, while "consisting of" is interpreted as exhaustive.

Table 2.6 illustrates the difference between the verbs "comprising" and "having" versus the word "consisting of" using an infringement example. The verbs "comprising" and "having" describe a device and a method in a non-exhaustive manner. This means that a claim formulation such as "method comprising steps 1 and 2" also covers a method that implements steps 1, 2, and an additional step 3. On the other hand, if the verb "consisting of" is used, this leads to an exhaustive enumeration of features, so that within the scope of protection of a claim formulation "device consisting of features 1 and 2", an embodiment with features 1, 2, and 3 is not included, and the patent proprietor therefore has no right to injunctive relief against this embodiment.

It is therefore generally advisable to enumerate the individual features of a claim using the verbs "comprising", "having", or the participles "comprising", "having", in order to claim a broad scope of protection. The verb "consisting of" should be avoided or used deliberately to describe an exhaustive enumeration. An exhaustive description can be used in particular to distinguish from the prior art and to make it clear that the advantageous effect of the invention lies in the fact that additional features are not required and the inventive effect is nevertheless achieved.

---

[68] Article 84, sentence 2 EPC; EPO, https://www.epo.org/de/legal/case-law/2022/clr_ii_a_5_1.html, accessed on 1.7.2025.

[69] Section 14, sentence 2 Patent Act.

Table 2.6 Scopes of protection: “comprising” and “having” versus “consisting of”

	comprising (non-exhaustive)	having (non-exhaustive)	consisting of (exhaustive)
Infringing embodiment	Claim 1: Device comprises features 1 and 2	Claim 2: Device has features 1 and 2	Claim 3: Device consists of features 1 and 2
Embodiment 1: Device comprises feature 1	No infringement	No infringement	No infringement
Embodiment 2: Device comprises features 1 and 2	Infringement	Infringement	Infringement
Embodiment 3: Device comprises features 1, 2, and 3	Infringement	Infringement	No infringement
Embodiment 4: Device comprises features 1 and 3	No infringement	No infringement	No infringement

2.42 Disclaimer

A disclaimer constitutes an exception provision to exclude a part of the previous scope of protection of the claim. The claim is thereby limited by the inclusion of a negative feature.

A claim must be described by including the features that positively define the scope of protection. Disclaimers for the negative limitation of the scope of protection are only permissible if an alternative clear and concise positive formulation is not possible.[70]

If a disclaimer is included during the grant procedure or in opposition or nullity proceedings, it must be examined whether the subject matter of the claim thus limited is, from the perspective of a person skilled in the art, directly and unambiguously disclosed by the originally filed application documents.[71] Under no circumstances may a disclaimer result in an aliud, i.e., a different subject matter of the claim.

A disclaimer may be permissible to restore the novelty of a claim in view of prior art documents that are not suitable for assessing inventive step. This applies in particular to subsequently published patent applications that were filed before the own filing date and published only on or after the filing date.[72]

It is not permissible to exclude individual prior art documents from the scope of protection in a mosaic fashion by the stepwise inclusion of disclaimers. Such an approach to fulfill the novelty criterion of patent law is not accepted by the patent office. However, it is permissible to exclude so-called “accidental” anticipations from the scope of protection by means of disclaimers. “Accidental” anticipations are prior art documents that are thematically so far removed from the invention that they would not have been considered by

[70] EPO G 1/03 and T 4/80.

[71] EPO G 2/10.

[72] Section 3(2) Patent Act or Article 54(3) EPC.

a person skilled in the art.[73] Although accidental anticipations describe all features of the invention, the technical teaching of the anticipation is intended to achieve a completely different technical effect in a completely different technical field, so that the anticipation and the invention may (by chance) have the same features, but are otherwise entirely unrelated. However, it must be ensured that a disclaimer excludes only as much from the scope of protection as is absolutely necessary to restore novelty in light of the accidental anticipation.[74]

The inclusion of a disclaimer that is not contained in the originally disclosed application documents must under no circumstances result in an additional technical contribution that is used to assess inventive step.[75]

A disclaimer can be particularly useful to comply with patentability exclusions without having to give up too much scope of protection.[76]

Except in extreme exceptional cases, it is therefore not permissible to establish novelty with disclaimers that were not disclosed in the originally filed application documents. In particular, it is not permissible to use disclaimers not originally disclosed to establish inventive step and enablement of a claim. Otherwise, the function of the patentability criteria would be undermined by selectively cutting out the scope of protection of the invention. Moreover, the use of disclaimers not originally disclosed would constitute an inadmissible extension.

2.43 Practical Examples

Examples from practice are presented to introduce specific claim formulations that illustrate the theoretical principles described.

Example 1: Cable Holder for Mounting on a Wall

Utility model publication DE 20 2017 106 959 U1 describes a device for detachably fixing loose cables to a wall.[77] The utility model publication provides the following explanation:

> *"One idea of the invention can be seen as providing a device that secures loose cable sections. This can in particular prevent accidents caused by tripping over loose cables lying around. For this purpose, a support or base is provided, on which projections, for example in the form of cylinders, are arranged. Cables can be wound around the bodies of the cylinders, thereby securing loose cable sections."*[78]

---

[73] EPO G 1/03, 4.4.2004, OJ EPO 2004, 413—Disclaimer/PPG.

[74] EPO G 1/03, 4.4.2004, OJ EPO 2004, 413.

[75] EPO, https://www.epo.org/de/legal/guidelines-epc/2025/h_v_4_2_1.html, accessed on 1.7.2025.

[76] EPO T 19/90, 3.10.1990, OJ EPO 1990. 476—Oncomouse/Harvard.

[77] DPMA, https://depatisnet.dpma.de/DepatisNet/depatisnet?action=pdf&docid=DE202017106959U1&xxxfull=1, accessed on 10.6.2025.

[78] DPMA, https://depatisnet.dpma.de/DepatisNet/depatisnet?action=pdf&docid=DE202017106959U1&xxxfull=1, accessed on 10.6.2025.

Fig. 2.8 Figs. 1 to 3 of DE202017106959U1[79]

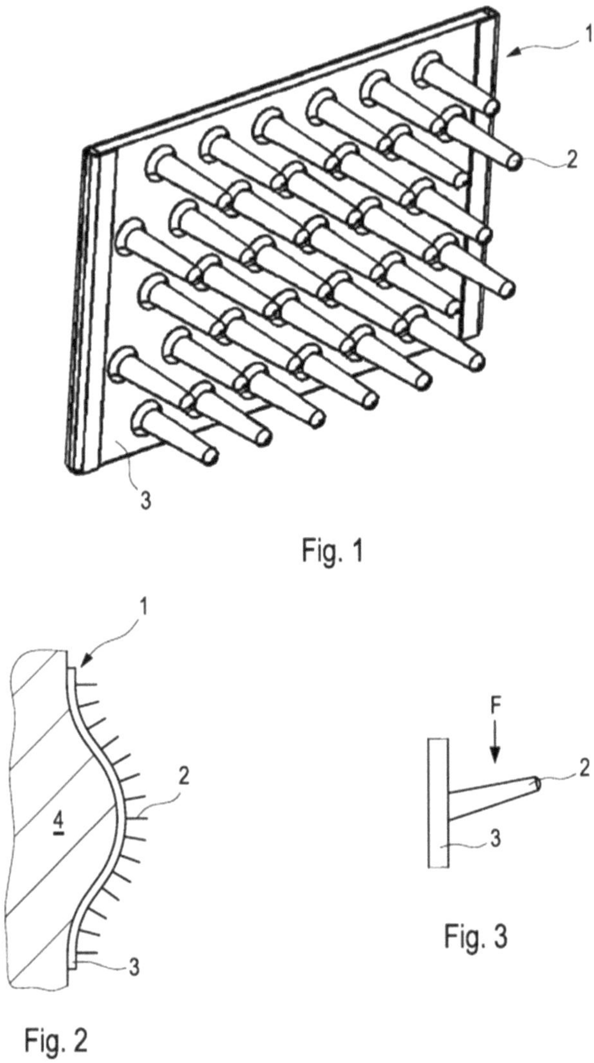

Figure 2.8 shows a mat 1 with a multitude of knobs or projections 2, which can be wrapped with cable sections to secure them. The mat 1 can in particular be mounted on a wall.

The invention is intended to create a “tidy” workplace situation by removing loose cables from the floor. This eliminates the risk of accidents caused by getting tangled in a mess of cables.

---

[79] DPMA, https://depatisnet.dpma.de/DepatisNet/depatisnet?action=pdf&docid=DE202017106959U1&xxxfull=1, accessed on 10.6.2025.

The protection claims are as follows; for utility models, the term "protection claims" is used instead of "patent claims":

"1. Device for holding a cable or a hose, wherein the device comprises:
a projection (2) for winding a cable and
a base (3) for holding the projection (2),
characterized in that
the base (3) can be arranged flat on any surface.

2. Device according to claim 1, characterized in that the device comprises 2, 3, 4, 5, 6, 7, 8, 9 or any number of projections (2).

3. Device according to one of claims 1 or 2, characterized in that the projections (2) are formed as cylinders, pyramids, cones, or spikes.

4. Device according to any of the preceding claims, characterized in that the device is suitable for mounting on a house wall (4).

5. Device according to any of the preceding claims, characterized in that the device comprises a holder (1) to which the base (3) can be detachably or non-detachably arranged.

6. Device according to any of the preceding claims, characterized in that the projections (2) are formed with a spherical or thickened shape at the outer end.

7. Device according to any of the preceding claims, characterized in that the projections (2) are arranged relative to each other such that a cable can just be wound around them due to the stiffness of its material.

8. Device according to any of the preceding claims, characterized in that the base (3) has a first area with first projections and a second area with second projections, wherein the first projections have different labeling and/or color from the second projections.

9. Device according to claim 8, characterized in that the first projections have a different spacing from each other than the second projections have from each other.

10. Device according to any of the preceding claims, characterized in that the base (3) and the projections (2) are formed in one piece and/or wherein the base (3) and/or the projections (2) consist of a plastic and/or an elastic material and/or elastomer."[80]

Example 2: Kink Holder
DE 20 2017 106 960 U1 describes a flexible cable holder.[81] The utility model specification reveals the essential essence of the invention:

---

[80] DPMA, https://depatisnet.dpma.de/DepatisNet/depatisnet?action=pdf&docid=DE202017106959U1&xxxfull=1, accessed on 10.6.2025.

[81] DPMA, https://depatisnet.dpma.de/DepatisNet/depatisnet?action=pdf&docid=DE202017106960U1&xxxfull=1, accessed on 10.6.2025.

Fig. 2.9 Figs. 1 and 2 of DE2 02017106960U1[83]

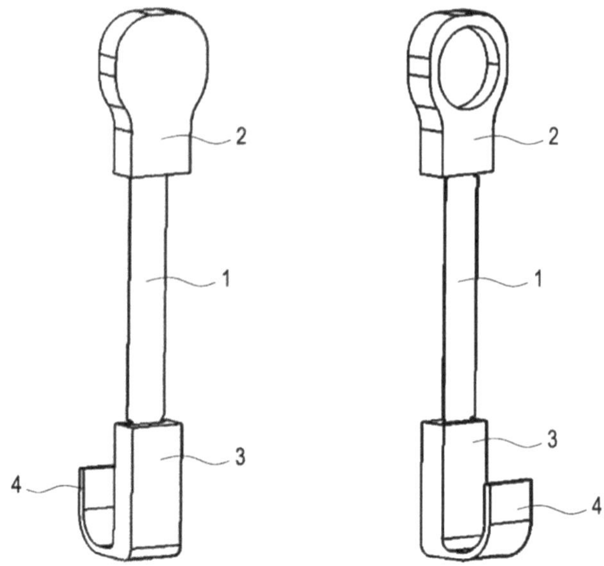

> *"An idea of the invention can be seen as providing a device that can bundle several cables into a strand, similar to a cable with multiple wires. In this case, the device is designed as a type of cable tie, whereby the cable holder can be easily released. This makes the device particularly suitable as an organizing system for cables, where the cables are only to be arranged in this way for a limited period of time."*[82]

Figure 2.9 shows an inventive bend holder, which comprises a flexible band 1 that can be bent in such a way that two magnets or a magnet and a ferromagnetic element can be placed on top of each other, resulting in a detachable connection. This creates a loop into which a cable section can be placed.

Figure 2.10 shows the bend holder, which is attached to a wall 8 with an element 5.

Figure 2.11 shows the folded bend holder, where the metal band 1 is bent in such a way that a loop is formed in which a cable 9 can be accommodated.

The claim set of the utility model reads as follows:

---

[82] DPMA, https://depatisnet.dpma.de/DepatisNet/depatisnet?action=pdf&docid=DE202017106960U1&xxxfull=1, accessed on 10.6.2025.

[83] DPMA, https://depatisnet.dpma.de/DepatisNet/depatisnet?action=pdf&docid=DE202017106960U1&xxxfull=1, accessed on 10.6.2025.

Fig. 2.10 Fig. 3 of DE202017 106960U1[84]

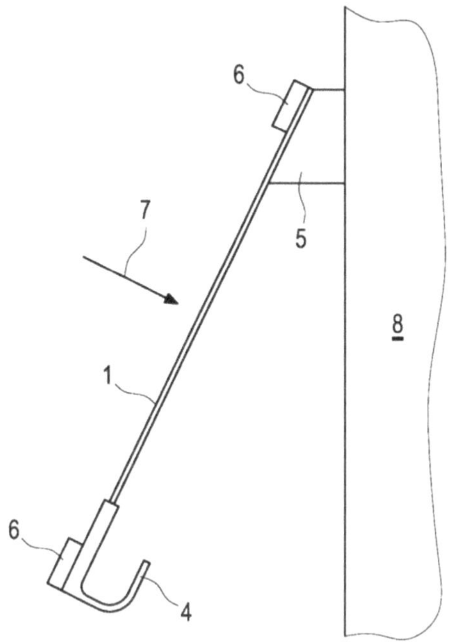

Fig. 3

"1. Device for holding a cable (9) or hose, wherein the device comprises:
a first end section (2),
a second end section (3) for connecting to the first end section (2), and
a middle section (1) for at least partially enclosing the cable (9) or hose, wherein the middle section (1) is arranged between the first and the second end section (2, 3),
characterized in that
the first end section (2) can be connected to the second end section (3), wherein the middle section (1) is bendable around the cable (9) or the hose.

2. Device according to claim 1, characterized in that the first end section (2) is a first magnet and the second end section (3) is a metal or a second magnet to which the first magnet can adhere.

3. Device according to one of claims 1 or 2, characterized in that the middle section (1) is a metal strip or a plastic strip, wherein the middle section is elongated in a first arrangement and bent in a second arrangement.

[84] DPMA, https://depatisnet.dpma.de/DepatisNet/depatisnet?action=pdf&docid=DE202017106960U1&xxxfull=1, accessed on 10.6.2025.

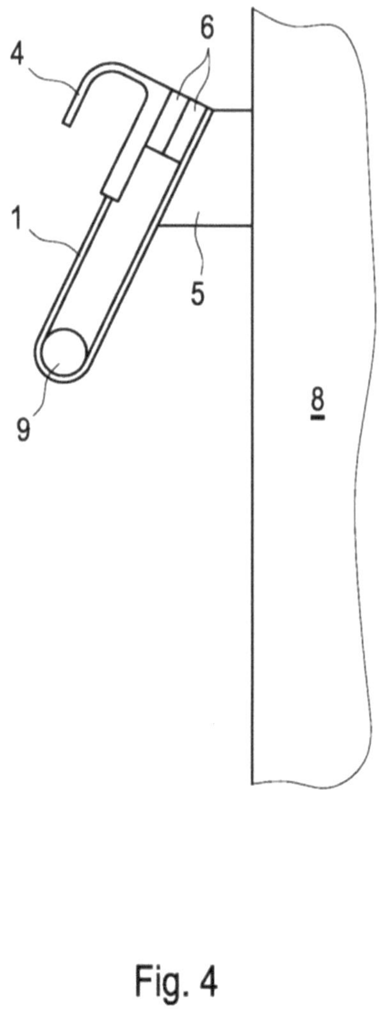

Fig. 2.11 Fig. 4 of DE202017 106960U1[85]

4. Device according to claim 3, characterized in that the metal strip has a first cross-section in the elongated arrangement and/or wherein the metal strip has a second cross-section in the bent arrangement, which is different from the first cross-section.

5. Device according to claim 4, characterized in that the first cross-section is not straight and/ or wherein the second cross-section is approximately straight.

[85] DPMA, https://depatisnet.dpma.de/DepatisNet/depatisnet?action=pdf&docid=DE202017106960U1&xxxfull=1, accessed on 10.6.2025.

6. Device according to any of the preceding claims, characterized in that the device comprises a hook (4) for elongating the bent middle section (1) and/or wherein the hook (4) is arranged on the first or the second end section (2, 3).

7. Device according to any of the preceding claims, characterized in that the first end section (2) can be detachably or non-detachably connected to the second end section (3) and/or wherein the first end section (2) is designed to latch with the second end section (3) and/or wherein the first end section (2) can latch into a recess of the second end section (3) if the middle section (1) is bent.

8. Device according to any of the preceding claims, characterized in that the device comprises a stopper (6) for damping an impact of the first end section (2) on the second end section (3) when the middle section (1) is bent.

9. Device according to any of the preceding claims, characterized in that the device comprises a support (5) for angularly aligning the device relative to a wall (8) or a vertical surface.

10. Device according to any of the preceding claims, characterized in that the device is formed in one piece, in particular as a component made of strip steel, and/or wherein the device comprises a holder for attachment to a wall or a vertical surface."[86]

The claim structure is shown in Fig. 2.12.

The claim structure in Fig. 2.12 shows that claims 4 and 5 refer only to their respective predecessors. This may be necessary, for example, because a term is first introduced in the preceding claim, which is then further specified.

Example 3: Mirror with Display

Patent DE 10 2015 104 437 A1 describes a mirror that integrates various display elements.[87] The essential essence of the invention can be taken from the patent as follows:

"As an idea of the invention, it can be considered to provide a mirror that enables the display of information/data while viewing oneself or during personal care. This information/data may relate to the health status of the person and/or include more general information (weather report, political, economic news, TV broadcasts, etc.)."[88]

Figures 2.13, 2.14, 2.15, and 2.16 show embodiments according to the invention. The meaning of the reference numerals can be found in the reference numeral list of the patent:

[86] DPMA, https://depatisnet.dpma.de/DepatisNet/depatisnet?action=pdf&docid=DE202017106960U1&xxxfull=1, retrieved on 25.6.2025.

[87] DPMA, https://depatisnet.dpma.de/DepatisNet/depatisnet?action=pdf&docid=DE102015104437A1&xxxfull=1, retrieved on 11.6.2025.

[88] DPMA, https://depatisnet.dpma.de/DepatisNet/depatisnet?action=pdf&docid=DE102015104437A1&xxxfull=1, retrieved on 11.6.2025.

Fig. 2.12 Claim structure of the device claims of DE202017106960U1

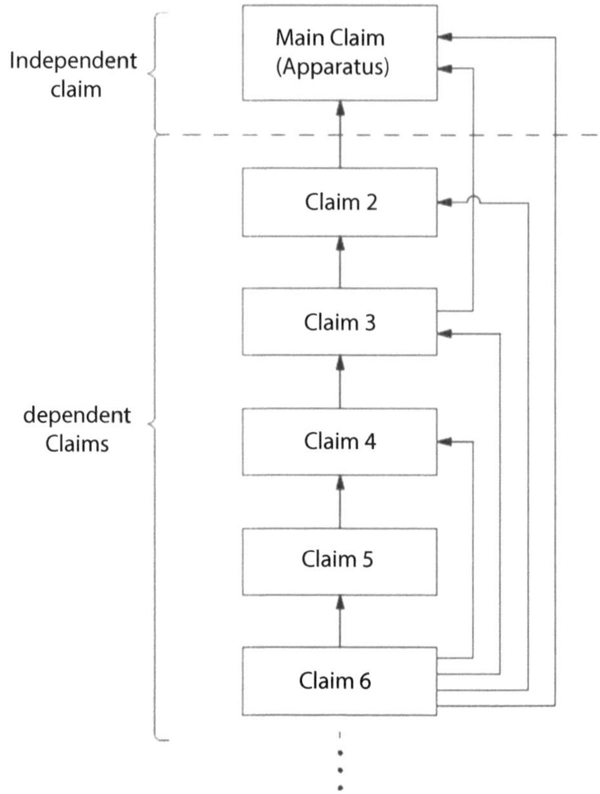

"LIST OF REFERENCE NUMERALS

1 *reflective surface*
2 *infrared spot*
3 *control unit/computer*
4 *Bluetooth*
5 *display*
6 *input field*
7 *cable*
8 *cable*
9 *person*
10 *line of sight*
11 *reflected radiation*
12 *emitted radiation*
13 *communication channel mirror with mobile phone/smartphone*
14 *communication channel mobile phone/smartphone with internet*
15 *communication channel computer/server with internet*
16 *internet*
17 *computer/server*
18 *mobile phone/smartphone*
19 *separating bar*

Fig. 2.13 Figs. 1 and 2 of DE102015104437A1[90]

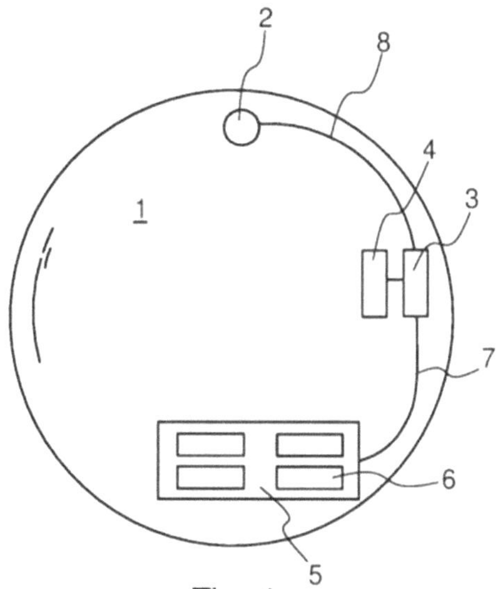

Fig. 1

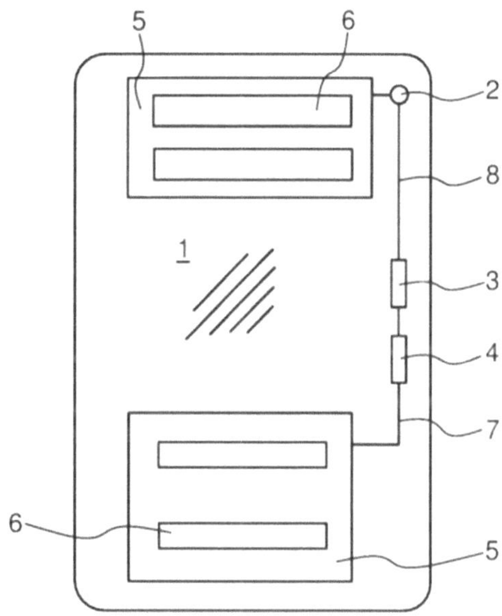

Fig. 2

[90] DPMA, https://depatisnet.dpma.de/DepatisNet/depatisnet?action=pdf&docid=DE102015104437A1&xxxfull=1, accessed on 11.6.2025.

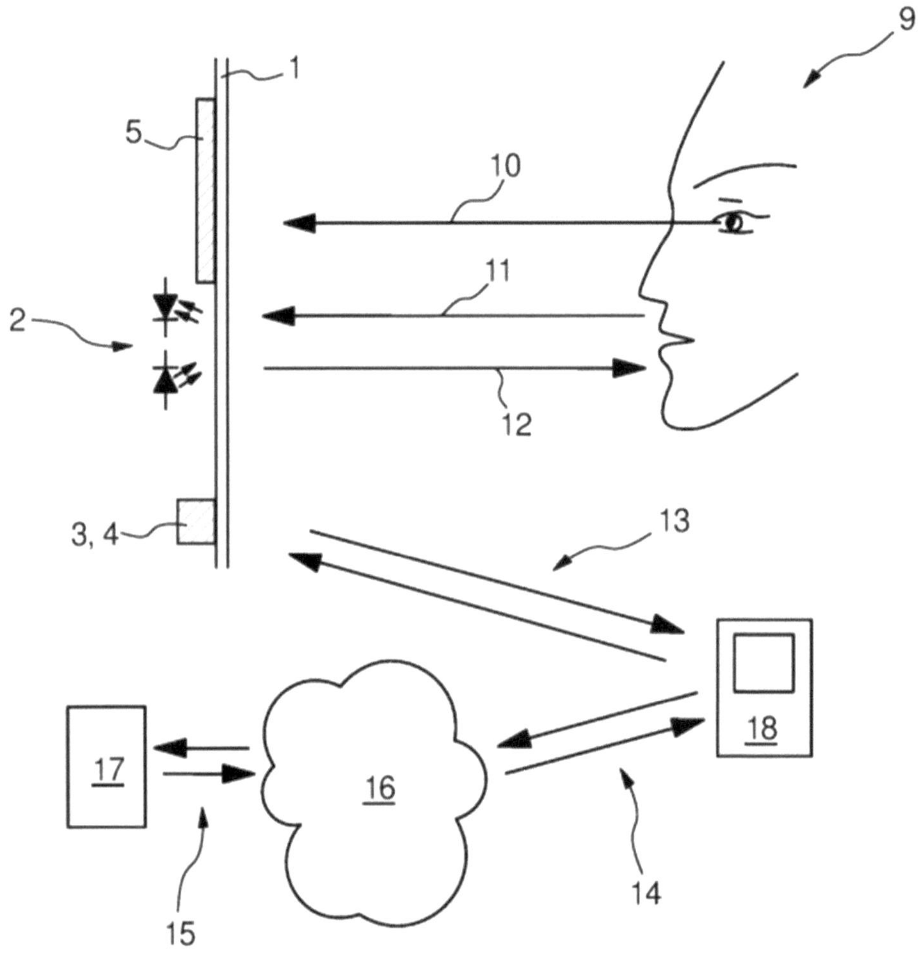

Fig. 2.14 Fig. 3 of DE102015104437A1[91]

20 *partially transparent section*
21 *partially transparent section*
22 *partially transparent section*
23 *partially transparent section"*[89]

[89] DPMA, https://depatisnet.dpma.de/DepatisNet/depatisnet?action=pdf&docid=DE102015104437A1&xxxfull=1, accessed on 11.6.2025.

[91] DPMA, https://depatisnet.dpma.de/DepatisNet/depatisnet?action=pdf&docid=DE102015104437A1&xxxfull=1, accessed on 11.6.2025.

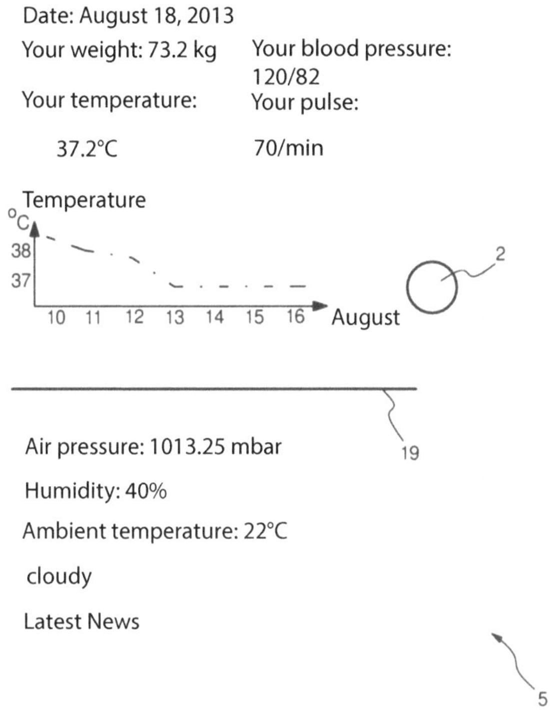

Fig. 4

Fig. 2.15 Fig. 4 of DE102015 104437A1[92]

Figure 2.13 shows in Fig. 1 a disk-shaped inventive mirror and in Fig. 2 a rectangular inventive mirror. Both mirrors contain displays 5 with input fields 6 for conveying information. In addition, a computer 3 for controlling the individual elements of the mirror is also shown.

Figure 2.14 shows the information paths that arise between the inventive mirror 1 and the observing person 9, as well as between the mirror 1 and a smartphone 18 or a computer 17. In addition, the mirror 1 contains sensors 2 that perform a health diagnosis of the person 9. The data from the mirror 1 can be sent to the smartphone 18 or to a computer 12, which, for example, is installed in the general practitioner's office of person 1.

Figure 2.15 shows a particular embodiment of the display 5 of the inventive mirror, in which health-related data (blood pressure, pulse, and weight) and general data (air pressure, humidity, and ambient temperature) are displayed.

[92] DPMA, https://depatisnet.dpma.de/DepatisNet/depatisnet?action=pdf&docid=DE102015104437A1&xxxfull=1, accessed on 11.6.2025.

Fig. 2.16 Figs. 4 and 5 of DE202017107510U1[93]

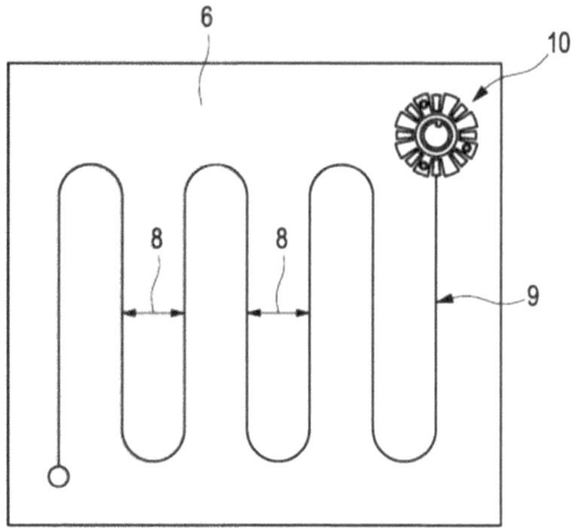

Fig. 4

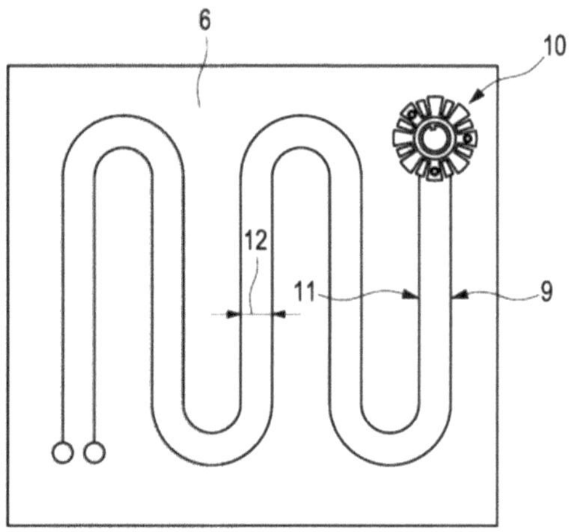

Fig. 5

[93] DPMA, https://depatisnet.dpma.de/DepatisNet/depatisnet?action=pdf&docid=DE202017107510U1&xxxfull=1, retrieved on 13.6.2025.

The claims of the patent read as follows:

"1. Mirror for reflecting a person, comprising:
a reflective surface (1), wherein the reflective surface (1) has a first section (21, 22, 23) that is at least partially transparent, and
a display (5) for displaying data,
characterized in that
the display (5) is arranged, from the perspective of the person, behind the first section (21, 22, 23), and wherein the data comprise information about the person's health, for example body temperature, blood pressure and/or pulse, and/or information about the environmental situation, for example air pressure, humidity, ambient temperature, current weather report, and/or general political, economic or other news and/or TV broadcasts, wherein the reflective surface (1) has a second section (20) that is at least partially transparent, and/or wherein the mirror comprises at least one sensor (2) for detecting the person and/or wherein the sensor (2) is arranged, from the perspective of the person, behind the second section (20), and wherein the mirror has an interface (4) to further devices, for example a blood pressure monitor, a heart rate monitor, a personal scale, and/or to the Internet, and/or wherein the interface (4) is a USB interface or a wireless interface.

2. Mirror according to claim 1, characterized in that the mirror comprises at least one device (2) for emitting radiation and/or for receiving radiation reflected by the person, and/or wherein the reflective surface (1) has a third section (20) that is at least partially transparent, and/or wherein the device (2) is arranged, from the perspective of the person, behind the third section (20).

3. Mirror according to claim 1 or 2, characterized in that the sensor (2) is selected from the group consisting of infrared sensor, radio sensor, image sensor

4. Mirror according to any of the preceding claims, characterized in that the display (5) and/or the sensor (2) and/or the device (2) are arranged within or behind the mirror.

5. Mirror according to any of the preceding claims, characterized in that the mirror comprises a control unit/a computer (3) for controlling the display (5), the sensor (2) and/or the device (2).

6. Mirror according to any of the preceding claims, characterized in that the first section (21, 22, 23) comprises the entire reflective surface (1) and/or wherein the first section (21, 22, 23) and/or the second section (20) and/or the third section (20) are of equal or different transparency."[94]

[94] DPMA, https://depatisnet.dpma.de/DepatisNet/depatisnet?action=pdf&docid=DE102015104437A1&xxxfull=1, accessed on 11.6.2025.

Example 4: Contact Points
Utility model DE 20 2017 107 510 describes an adhesive element for providing defined contact points. The utility model specification contains the following paragraph, which describes the essence of the invention:

> *"An idea of the invention can be seen in designing a connecting element with injection-molded plastic in such a way that contact possibilities arise with conductive tracks on the substrate on which the connecting element with injection-molded plastic is applied. In this way, an electrical analysis device can be used to check at the injection-molded plastic whether there is moisture on the substrate. In the event of water ingress, a short circuit occurs between the conductive tracks, which can be detected by the analysis device."*[95]

Figure 2.16 shows two embodiments of the invention, with Fig. 4 depicting a meandering conductor track 9 formed on a substrate 6. If moisture causes a short circuit 8 between individual sections of the conductor track 9, the resistance of the conductor track 9 decreases, thereby detecting the moisture.

Figure 5 of Fig. 2.16 shows two embodiments of the invention, with Fig. 4 depicting a meandering conductor track 9 formed on a substrate 6. If moisture causes a short circuit 8 between individual sections of the conductor track 9, the resistance of the conductor track 9 decreases, thereby detecting the moisture.

The set of claims of the utility model reads as follows:

> *"1. Connecting element (4) with injection-molded plastic (10) for detachable or non-detachable connection to a substrate (6), wherein the injection-molded plastic (10) comprises:*
> *a tab (3) for ensuring an electrical or galvanic connection between an electrical conductor on the substrate (6) and a connecting unit.*
>
> *2. Connecting element (4) according to claim 1, characterized in that the electrical conductor is a conductive track or a conductive layer or a conductive surface or a conductive coating (9, 11) and/or wherein the electrical conductor (9, 11) is at least partially coated onto the tab (3) and/or wherein the connecting unit is a plug.*
>
> *3. Connecting element (4) according to one of claims 1 or 2, characterized in that the electrical conductor is at least partially on the tab (3) or wherein the electrical conductor is at least partially beneath the tab (3).*
>
> *4. Connecting element according to claim 3, characterized in that the electrical conductor is at least partially beneath the tab (3) and the tab (3) has a recess (2) for the connecting unit to access the electrical conductor.*
>
> *5. Connecting element (4) according to claim 1, characterized in that the connecting element (4) comprises a devicefor orienting the connecting unit (4).*

[95] DPMA, https://depatisnet.dpma.de/DepatisNet/depatisnet?action=pdf&docid=DE202017107510U1&xxxfull=1, retrieved on 13.6.2025.

6. Connecting element (4) according to any of the preceding claims, characterized in that the injection-molded plastic (10) has a web (14) for preventing an electrical or galvanic connection of at least two tabs (3) and/or wherein one or both tabs (3) are recessed relative to the web (14).

7. Connecting element (4) according to any of the preceding claims, characterized in that the tab (3) is inclined relative to a substrate (6) to which it is to be connected

8. Connecting element (4) according to any of the preceding claims, characterized in that the injection-molded plastic (10) has 2, 3, 4, 5, 6, 7, 8, 9, 10 or any number of tabs (3) and/or wherein the injection-molded plastic (10) comprises 2, 3, 4, 5, 6, 7, 8, 9, 10 or any number of webs (14).

9. Connecting element (4) according to any of the preceding claims, characterized in that the injection-molded plastic (10) comprises 1, 2, 3, 4, 5, 6 or any number of recesses (2) for fixing a mating plug."[96]

The structure of the set of claims is shown in Fig. 2.17.

Example 5: Blood Pressure Monitor

Patent application DE 10 2015 104 432 A1 describes a blood pressure monitor for long-term blood pressure measurement. The patent specification states:

"An idea of the invention can be seen as providing a device that enables blood pressure measurement and is capable of transmitting its measurement results to a watch for display. This allows the device to be made compact, consisting of only a single cuff, and eliminates the need for a dedicated display unit. The cuff can be worn on the upper arm or the wrist."[97]

Fig. 2.18 shows a cuff 1 for blood pressure measurement, which transmits its results to a display element 2 in the form of a watch.

Fig. 2.19 shows a network of elements for monitoring a person's blood pressure, where the measurement results from cuff 1 are transmitted to a watch 2, to a mobile phone/smartphone 3, and via the Internet 4 to a computer 5 in a hospital or doctor's office.

The set of claims of the patent application reads as follows:

"1. System for measuring the blood pressure of a user, comprising
- *a cuff device (1) attachable to an upper arm of a user, with a sensor for measuring blood pressure and a data transmission device for transmitting the measured data associated with the user's blood pressure;*
- *a display device attachable to a wrist of the user, with a data receiving device for receiving the data transmitted by the data transmission device, wherein the data are transmitted wirelessly from the data transmission device to the data receiving device.*

[96] DPMA, https://depatisnet.dpma.de/DepatisNet/depatisnet?action=pdf&docid=DE202017107510U1&xxxfull=1, retrieved on 25.6.2025.

[97] DPMA, https://depatisnet.dpma.de/DepatisNet/depatisnet?action=pdf&docid=DE102015104432A1&xxxfull=1, accessed on 13.6.2025.

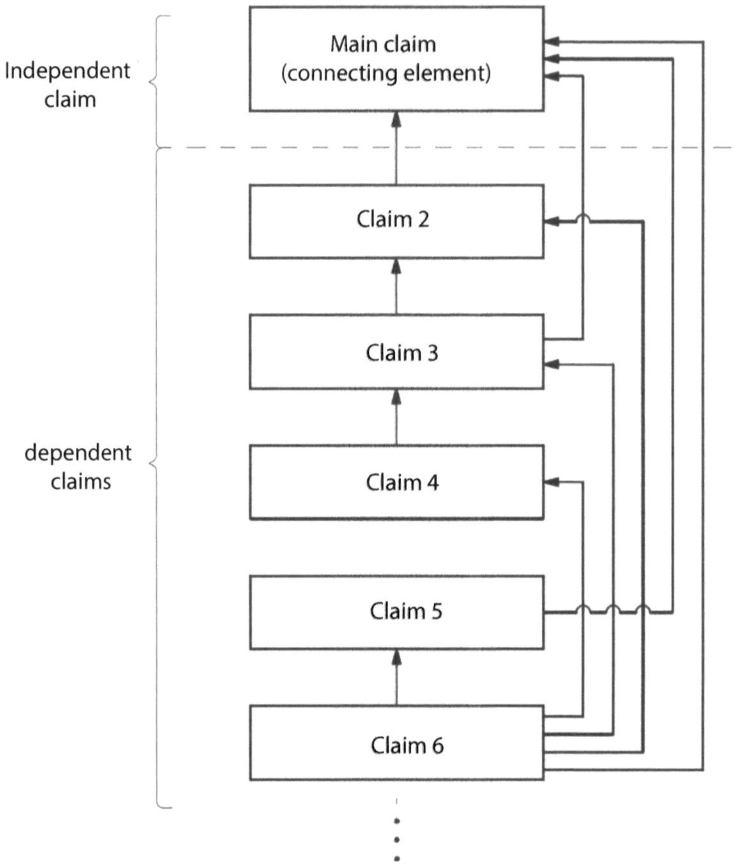

Fig. 2.17 Claim structure of the device claims of DE202017107510U1

2. System according to claim 1, wherein the display device comprises a wristband with a display field.

3. System according to claim 2, wherein the display field is arranged in a housing, in particular the housing of a watch or pulse watch.

4. System according to any of the preceding claims, wherein the display device comprises a motion detection device for recording the user's activity.

5. System according to any of the preceding claims, wherein the display device comprises at least one further sensor for detecting body signals of the user and/or the position of the user."[100]

---

[100] DPMA, https://depatisnet.dpma.de/DepatisNet/depatisnet?action=pdf&docid=DE102015104432A1&xxxfull=1, accessed on 25.6.2025.

Fig. 2.18 Fig. 1 of DE102015 104432A1[98]

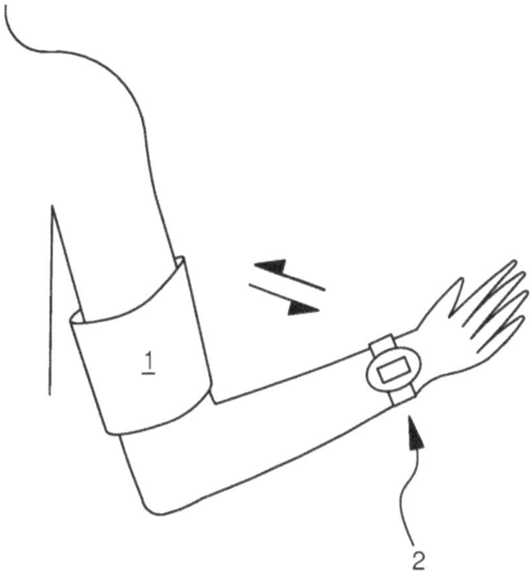

Fig. 1

Example 6: Washing Machine

Patent WO 2022/008447 describes a washing machine or dishwasher for cleaning laundry and dishes. The patent specification states:

> *"An idea of the invention can be seen as providing a washing chamber for laundry that is fully illuminated with ultraviolet radiation, in particular with UV-C radiation. In a particular embodiment, the ultraviolet radiation source interacts with a reflector so that ultraviolet radiation can be directed onto the laundry from any location within the washing chamber. Additionally or alternatively, the ultraviolet radiation source can be designed to be movable, in particular rotatable."*[101]

The essential core of the invention is therefore that the laundry is exposed to ultraviolet radiation in order to kill viruses, harmful germs, and bacteria.

Fig. 2.20 shows in Fig. 1 a washing machine 1 with a porthole 7, which is kept closed over a washing chamber 8 by a lock 6. In Fig. 2, circular radiation sources 2 are indicated, which emit disinfecting ultraviolet light into the washing chamber.

---

[98] DPMA, https://depatisnet.dpma.de/DepatisNet/depatisnet?action=pdf&docid=DE102015104432A1&xxxfull=1, accessed on 13.6.2025.

[101] DPMA, https://depatisnet.dpma.de/DepatisNet/depatisnet?action=pdf&docid=WO002022008447A1&xxxfull=1, accessed on 13.6.2025.

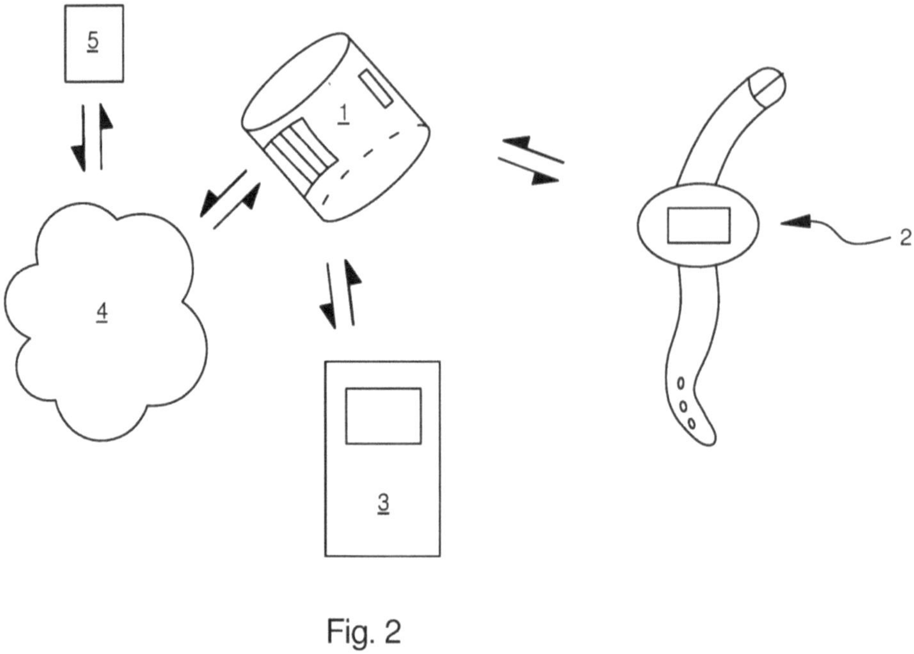

Fig. 2.19 Fig. 2 of DE102015104432A1[99]

Fig. 2.21 shows in Fig. 3 a meander-shaped configuration of an ultraviolet radiation source 2 inside the washing chamber 8. Fig. 4 shows another variant of a meander-shaped radiation source 2.

Fig. 2.22 shows in Fig. 5 a circular ultraviolet radiation source 2 and in Fig. 6 a cross-section of a washing machine according to the invention. Fig. 6 shows the radiation source 2, which is arranged behind the porthole, with a material 3 placed in front of the radiation source 2 that absorbs the ultraviolet radiation from the radiation source 2, so that the radiation cannot leave the washing chamber 8. A transparent wall 3 protects the radiation source from the wash water in the washing chamber 8 and prevents the laundry from getting caught in the radiation source 2.

Fig. 2.23 shows in Fig. 7 a variant of the washing machine with a radiation source 2 arranged in a circle around the porthole and a transparent wall 3 that keeps the laundry away from the radiation source. Fig. 8 shows a dishwasher with a radiation source 2, so that the dishes can be disinfected with ultraviolet radiation.

[99] DPMA, https://depatisnet.dpma.de/DepatisNet/depatisnet?action=pdf&docid=DE102015104432A1&xxxfull=1, accessed on 13.6.2025.

Fig. 2.20 Figs. 1 and 2 of WO002022008447A1[102]

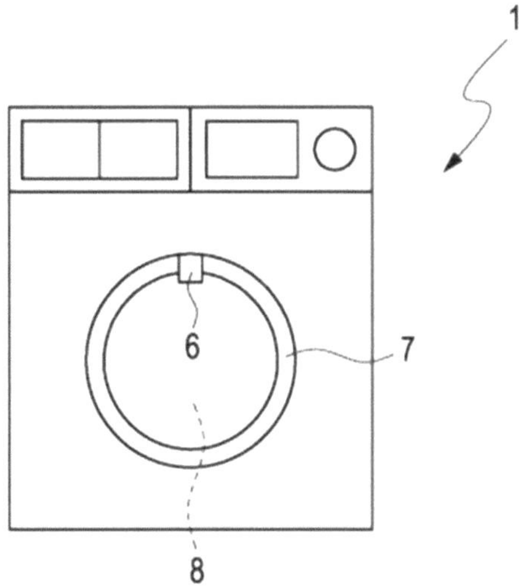

Fig. 1

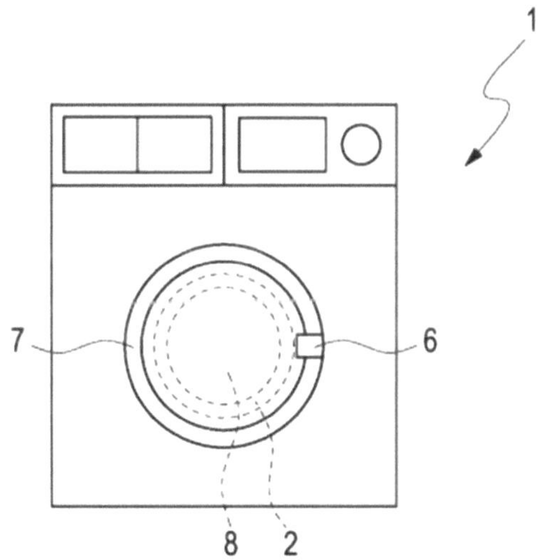

Fig. 2

[102] DPMA, https://depatisnet.dpma.de/DepatisNet/depatisnet?action=pdf&docid=WO002022008447A1&xxxfull=1, accessed on 13.6.2025.

Fig. 2.21 Figs. 3 and 4 of WO002022008447A1[103]

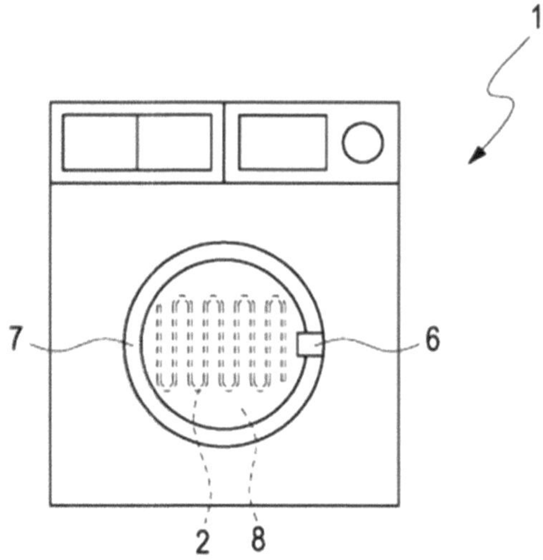

Fig. 3

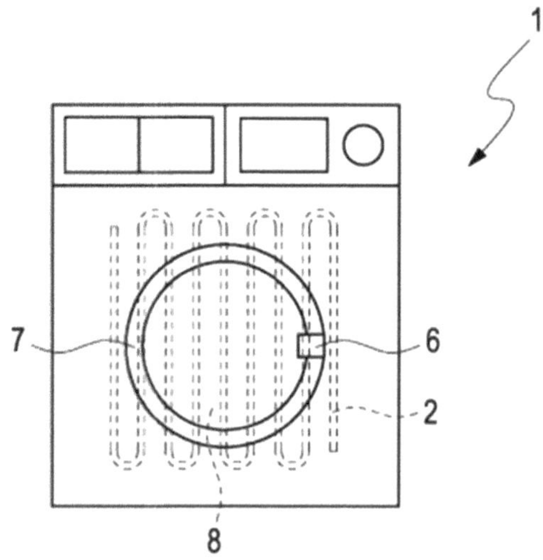

Fig. 4

[103] DPMA, https://depatisnet.dpma.de/DepatisNet/depatisnet?action=pdf&docid=WO002022008447A1&xxxfull=1, accessed on 13.6.2025.

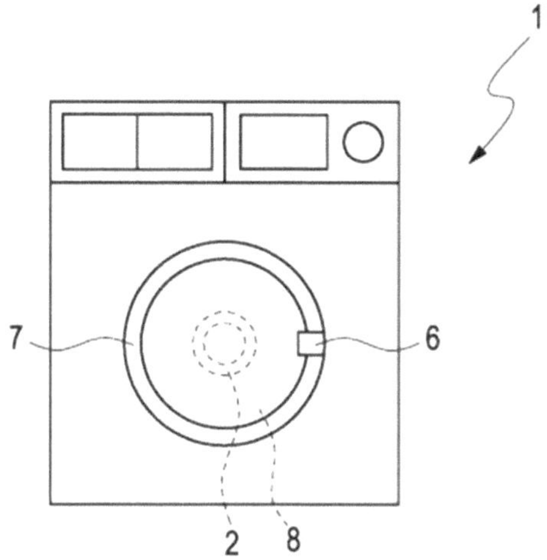

Fig. 5

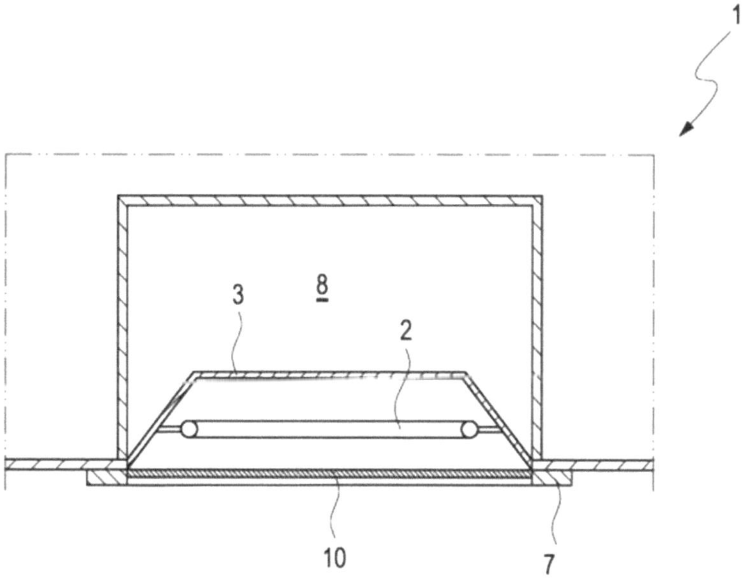

Fig. 6

Fig. 2.22 Figs. 5 and 6 of WO002022008447A1[104]

[104] DPMA, https://depatisnet.dpma.de/DepatisNet/depatisnet?action=pdf&docid=WO002022008447A1&xxxfull=1, accessed on 13.6.2025.

Fig. 2.23 Figs. 7 and 8 of WO002022008447A1[105]

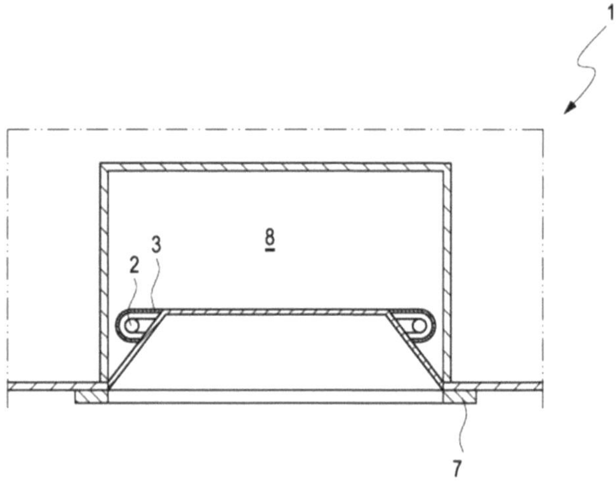

Fig. 7

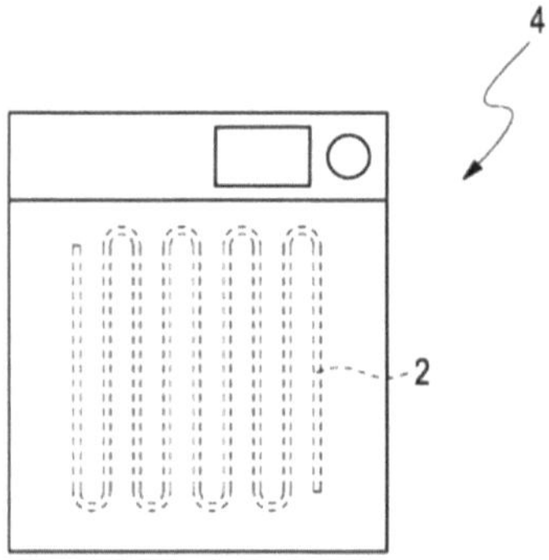

Fig. 8

The claims of the patent are as follows:

"1. Washing machine (1) or dishwasher (4) for cleaning laundry or dishes, wherein the washing machine (1) or the dishwasher (4) comprises a washing chamber (8) for the laundry or dishes, wherein the laundry or dishes are cleaned inside the washing chamber (8),

---

[105] DPMA, https://depatisnet.dpma.de/DepatisNet/depatisnet?action=pdf&docid=WO002022008447A1&xxxfull=1, accessed on 13.6.2025.

characterized in that
the washing machine (1) or the dishwasher (4) comprises an ultraviolet radiation source (2), in particular a UV-C radiation source, for irradiating the laundry or dishes, and wherein the ultraviolet radiation source (2) is arranged inside the washing chamber (8).

2. Washing machine (1) or dishwasher (4) according to claim 1, wherein the ultraviolet radiation source (2) is separated from the washing chamber (8) by a first material (3) that is permeable to ultraviolet radiation, in particular UV-C radiation, and/or wherein the material (3) is glass or plastic.

3. Washing machine (1) or dishwasher (4) according to claim 2, wherein the first material (3) is made of synthetic quartz glass.

4. Washing machine (1) or dishwasher (4) according to any of the preceding claims, wherein the ultraviolet radiation source (2) is designed to be movable, in particular rotatable, in order to scatter the ultraviolet radiation from the ultraviolet radiation source (2) arbitrarily within the washing chamber (8) during the washing process.

5. Washing machine (1) or dishwasher (4) according to any of the preceding claims, wherein the washing machine (1) or the dishwasher (4) comprises:

- *a first ultraviolet radiation source (2) that illuminates the washing chamber (8) from above, and/or*
- *a second ultraviolet radiation source (2) that illuminates the washing chamber (8) from below, and/or*
- *a third ultraviolet radiation source (2) that illuminates the washing chamber (8) from the left, and/or*
- *a fourth ultraviolet radiation source (2) that illuminates the washing chamber (8) from the right, and/or*
- *a fifth ultraviolet radiation source (2) that illuminates the washing chamber (8) from the front, and/or*
- *a sixth ultraviolet radiation source (2) that illuminates the washing chamber (8) from the rear.*

6. Washing machine (1) or dishwasher (4) according to any of the preceding claims, wherein a reflector is arranged in the washing chamber (1) so that ultraviolet radiation from the ultraviolet radiation source (2) can be reflected back into the washing chamber (8).

7. Washing machine (1) or dishwasher (4) according to claim 6, wherein the reflector is designed to be movable, in particular rotatable, in order to scatter the ultraviolet radiation arbitrarily during the washing process.

8. Washing machine (1) or dishwasher (4) according to any of the preceding claims, wherein the washing chamber (8) is lined with a second material that reflects ultraviolet radiation.

9. Washing machine (1) or dishwasher (4) according to any of the preceding claims, wherein the washing machine (1) or dishwasher (4) comprises a device (6) for detecting whether the washing chamber (8) is closed, and wherein the ultraviolet radiation source (2) can only be switched on if the device signals a closed washing chamber (8).

10. Washing machine (1) or dishwasher (4) according to any of the preceding claims, wherein the ultraviolet radiation source (2) is circular, semicircular, or meandering in shape and/or wherein a third material (10) that absorbs ultraviolet radiation is arranged between the exterior and the ultraviolet radiation source, and/or wherein the third material (10) is glass, perforated sheet metal, or plastic."[106]

Example 7: String Art Representation

The invention described in DE 10 2020 118 709 concerns the production of string art representations, in which nails are driven into a board and, depending on the density of the arrangement of the nails, a different visual appearance is achieved. Additionally, the nails can be connected with threads to enhance the visual effect, especially through the use of colorful threads.

Fig. 2.24 shows a device according to the invention for producing string art representations, in which, according to the invention, a continuous wire is first driven into a board and then cut off. This enables a very rapid creation of string art representations.

Fig. 2.25 shows an exemplary string art representation with nails 2, which are connected by threads 7 and thus create the visual impression of a star.

The set of claims of the patent reads as follows:

"1. Nail (2) for use in a string art representation, wherein the nail (2) does not have a thickened end in the form of a head.

2. Nail (2) according to claim 1, wherein the nail (2) has a tip at one end for driving into a plate (1).

3. Nail (2) according to one of claims 1 or 2, wherein the nail (2) has a flattened end to reduce the risk of injury.

4. Plate (1) as a string art representation, comprising:
one or a plurality of nails (2), according to one of the preceding claims.

5. Plate (1) according to claim 4, wherein the plate (1) further comprises:
a thread (7) connecting at least two nails (2) to each other.

6. Method for producing a plate (1) according to one of claims 4 or 5, comprising the steps of:
feeding a continuous wire (6),
driving the continuous wire (6) into the plate (1), and
cutting the continuous wire (6) to length, so that a nail (2) according to one of claims 1 to 3 is obtained.

[106] DPMA, https://depatisnet.dpma.de/DepatisNet/depatisnet?action=pdf&docid=WO002022008447A1&xxxfull=1, retrieved on 25.6.2025.

Fig. 2.24 Fig. 1 of DE102020118709[107]

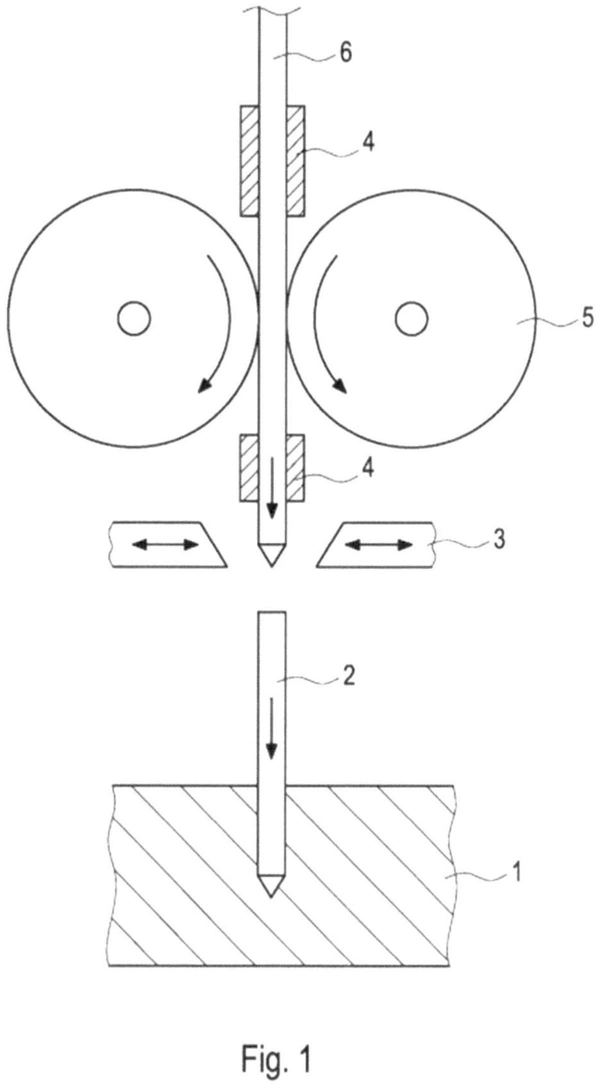

7. Method according to claim 6, wherein when cutting the continuous wire (6) to length, a tip is formed on the continuous wire (6) such that the continuous wire (6) can be more easily driven into the plate (1).

[107] DPMA, https://depatisnet.dpma.de/DepatisNet/depatisnet?action=pdf&docid=DE102020118709A1&xxxfull=1, accessed on 7/8/2025.

Fig. 2.25 Fig. 2 of DE102020118709[108]

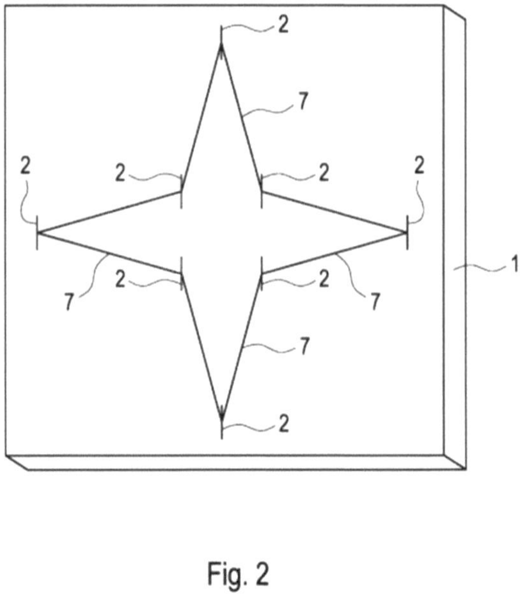

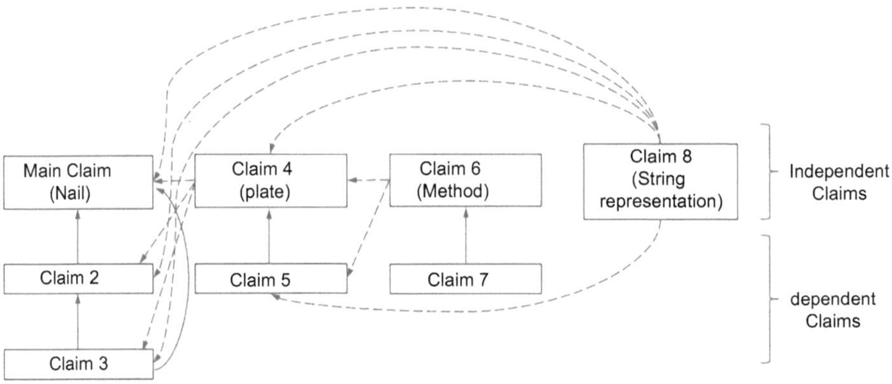

Fig. 2.26 Structure of the set of claims of DE102020118709A1

[108] DPMA, https://depatisnet.dpma.de/DepatisNet/depatisnet?action=pdf&docid=DE102020118709A1&xxxfull=1, accessed on 7/8/2025.

8. String art representation comprising:
at least two nails (2) according to one of claims 1 to 3, and
a plate according to one of claims 4 or 5."[109]

In Fig. 2.26 it can be seen that independent claims (claims 4, 6, and 8) can also contain references and thus include all features of the referenced claims. In Fig. 2.26, the additional reference of claim 6 to claims 1 to 3 has been omitted for clarity.[110]

[109] DPMA, https://depatisnet.dpma.de/DepatisNet/depatisnet?action=pdf&docid=DE102020118709A1&xxxfull=1, accessed on 7/8/2025.

[110] Section 9 paragraph 5 sentence 3 of the Patent Ordinance.

3 Interpretation of Claims

The interpretation of a claim serves to determine the correct understanding of the claim and, in particular, the scope of the protection described. For this purpose, different methods of interpretation have been developed, which are essentially derived from the interpretative approaches to legal norms in jurisprudence.

The interpretation of a legal norm is necessary to ensure the proper application of the law. Especially in the case of unclear and incomplete legal norms, only through interpretation can the relevant legal norm be correctly applied to a specific case. Jurisprudence has developed several variants for interpreting legal norms, which are initially applied individually and then considered in combination. In particular, grammatical, systematic, historical, and teleological interpretations have been developed. From these, the literal, systematic, teleological, and functional interpretations have been derived for the interpretation of patent claims. The historical interpretation is not applicable to patent claims.

3.1 Purpose of Interpretation

The interpretation of claims must satisfy two criteria. On the one hand, the patent holder should be granted appropriate patent protection, which results from the technical teaching of the patent. On the other hand, interpretation must provide legal certainty for the public by ensuring that the scope of protection of the patent claim can be determined clearly and unambiguously.

The patent holder must be granted the reward for their effort. The patented technical teaching must therefore be protected against any infringing imitation. The scope of protection granted by the courts will mirror the enrichment of the state of the art by the patented technical teaching. If the patented invention achieves a significant technological advance, the patent should be granted a broad scope of protection, insofar as the wording

T. H. Meitinger, *Drafting, Interpretation, and Amendment of Patent Claims*,
https://doi.org/10.1007/978-3-662-73622-7_3

of the claim allows. However, the specific effort or the extent of the work required to achieve the technical teaching of the patented invention is not to be taken into account. Even a "fortunate coincidence" can justify a broad scope of protection.

Patent claims are intended to ensure legal certainty for the public, so that any third party can determine the extent of the claimed embodiments directly and unambiguously through interpretation of the claims. It must be prevented that the public is surprised by a scope of protection that they would not have expected based on a reasonable interpretation of the claim wording.

3.2 Interpretation is a Question of Law

The question of the disclosure of a patented technical teaching, and thus whether the disclosure of the originally filed application documents or the granted scope of protection has been exceeded by an inadmissible extension, is a question of law and not a question of fact.[1] An inadmissible extension therefore cannot be decided by a technical expert. However, a technical expert may be consulted to establish a factual basis for the legal interpretation of claims.[2]

Since interpretation is a question of law and not a question of fact, the determination of an inadmissible extension or an extension of the scope of protection is, in principle, always subject to legal remedies, and, if the other requirements are met, may even be reviewed by way of appeal.[3]

3.3 Normative Character of the Claim

Granted patent claims have the character of legal norms.[4] Pursuant to a patent claim, legally regulated consequences can be enforced. It is therefore necessary that the interpretation of a patent claim is carried out according to objective criteria in order to maintain legal certainty.[5] For interpretation, an objective instance is determined, which is represented by the notional figure of the person skilled in the art in patent law.

According to Section 139, Paragraph 1, Sentences 1 and 2 of the Patent Act, there is a right to injunctive relief against the unauthorized user of a patented invention. The right to injunctive relief is vested in the patent holder. Granted patent claims therefore have the

[1] BGH 13.2.2007, X ZR 74/05—Kettenanordnung.
[2] BGH 11.10.2005, X ZR 76/04—Seitenspiegel.
[3] BGH 13.2.2007, X ZR 74/05—Kettenanordnung.
[4] BGH 11.10.2005 X ZR 76/04—Seitenspiegel; BGH 8.7.2008 X ZB 13/06—Momentanpol II.
[5] BGH 8.7.2008 X ZB 13/06—Momentanpol II.

character of an enforceable legal norm. However, the same does not apply to claims of a patent application that have not yet been granted.[6]

3.4 Uniformity of Interpretation

Patent claims must be interpreted uniformly. This means that the legal situation must not influence the interpretation.[7] In particular, it is not permissible to assume a narrow interpretation in opposition or nullity proceedings and a broad interpretation in infringement proceedings, which would favor the patent holder, or vice versa, which would be welcomed by an alleged infringer or opponent or nullity plaintiff.

In Germany, the principle of separation applies, according to which different courts are responsible for patent infringement and for the validity of the respective underlying patent. Furthermore, following an infringement action, an attack on the respective patent is regularly brought before the Federal Patent Court, so it is not uncommon for different courts to have to decide on the interpretation of the same patent at the same time.

The interpretation of a claim is a question of law, so the court seized must independently clarify the matter. It is not permissible for an expert to undertake the interpretation of a claim by preparing an expert opinion.[8] The regional courts and the Federal Patent Court are therefore obliged to interpret the claims independently. However, in order to strive for uniformity, it is customary for the courts to consider the submissions on the interpretation of the claims in the respective parallel proceedings.[9]

3.5 Limits of Interpretation

Interpretation is carried out from the perspective of a person skilled in the art, who also "reads in" features that are not expressly contained in the claim. Features can only be considered as "read in" if the skilled person would immediately also think of these

[6] BGH 11.10.2005, X ZR 76/04—Seitenspiegel; BGH 8.7.2008, X ZB 13/06—Momentanpol II, Einsele, Drafting of Patent Claims—Clear or Not Clear, Mitteilungen der deutschen Patentanwälte, 2014, 249, 254.

[7] GrBK EPO 4.4.2004, G 1/03, OJ EPO 2004, 413; GrBK EPO 30.8.2011, G 2/10, OJ EPO 2012, 367—Disclaimer/SCRIPPS; BGH 19.5.1981, X ZB 19/80, GRUR 1981, 812—Labelling Machine; Müller, Permissible Extension and Limitation within the Framework of Disclosure, Mitteilungen der deutschen Patentanwälte 1991, 10, 11.

[8] BGH 17 April 2007—X ZR 1/05—Pump Device; BGH 2 June 2015—X ZR 103/13—Crossbar; BGH 31 March 2009—X ZR 95/05—Road Construction Machine; BGH 29 June 2010—X ZR 193/03, BGHZ 186, 90 para. 15—Crimping Tool III.

[9] BGH 15.4.2010—Xa ZB 10(09 GRUR 2010, 950—Roller Forming Machine.

"read-in" features. Therefore, everything that is immediately and unambiguously apparent to the skilled person from the wording of the claim must be taken into account in the interpretation. However, this does not mean that everything that is obvious is to be included in the meaning of a claim. It is therefore necessary to distinguish precisely between what is considered "immediately and unambiguously" disclosed and what is obvious but cannot be attributed to the wording of the claim.[10]

3.6 Direct and Unambiguous Disclosure

The formulation "direct and unambiguous" disclosure has been developed by the European Patent Office in several decisions of the Enlarged Board of Appeal, with the formulation being referred to as the "gold standard" for determining whether there is an inadmissible extension.[11]

A "direct" disclosure is characterized by the fact that the relevant feature can be derived "without circumlocution" from the application documents. "Without circumlocution" can exclude features that are merely obvious. If something is obvious, it does not automatically mean that the skilled person would derive it "directly" or "without circumlocution" from the patent specification.[12]

A disclosure is "unambiguous" if the relevant feature is consistent with the entire technical teaching of the invention in such a way that there can be no doubt that precisely this feature is meant. However, if alternatives arise, there is no unambiguous disclosure.[13]

3.7 Skilled Person for Interpretation

Interpretation of patent claims must be carried out according to objective and comprehensible principles. Patent claims are to be interpreted from the perspective of the skilled person, who is assumed to have a technically correct understanding. In particular, the skilled person is able to recognize technically nonsensical embodiments that result from the claim wording. Technically nonsensical embodiments should not be included within the scope of protection.

[10] BGH 18.3.2014, X ZR 77/12—Protein Separation.

[11] Enlarged Board of Appeal EPO 19.11.1992, G 3/89, OJ EPO 1993, 117; Enlarged Board of Appeal EPO 19.11.1992, G 11/91, OJ EPO 1993, 125.

[12] EPO T 6/84, OJ EPO 1985, 238.

[13] German Federal Court of Justice 7.12.1978, X ZR 4/76—Aufhänger.

The skilled person in patent law is the authority for understanding a claim. In particular, the skilled person is required to decide which embodiments are to be assigned to the scope of protection of a claim and which no longer belong to the scope of protection. In particular, the skilled person will not consider technically nonsensical embodiments as intended embodiments.

3.8 Literal Interpretation

In literal or word-for-word interpretation, the wording is analyzed and the meaning of the individual terms in both everyday and technical language is explored. The totality of the meanings of the terms of a claim results in the literal interpretation of the claim. Literal interpretation forms the starting point for any interpretation of a claim.

3.9 Systematic Interpretation

Systematic interpretation fulfills the requirement of the legislator, who in Section 14, sentence 2 of the Patent Act stipulates that the description and drawings of a patent specification must be taken into account when interpreting the claims. In systematic interpretation, the patent claim is interpreted in the context of the entire disclosure of the patent specification. Article 69(1), sentence 2 EPC prescribes the application of systematic interpretation for European patents.

3.10 Teleological Interpretation

Legal science uses teleological interpretation to search for the meaning and purpose (ratio legis) of a legal norm. From a patent law perspective, teleological interpretation is used in particular to determine what task an invention is intended to fulfill. Teleological interpretation seeks the underlying idea behind the invention and does not adhere to the wording of the relevant claim. Teleological interpretation can lead to uncertainties, as the objective of the inventive teaching may change. In particular, the applicant is permitted to adapt the technical problem described in the patent application to newly researched prior art, so that the original objective may no longer be relevant.[14]

[14] Osterrieth, Part 6. Patent Infringement in Osterrieth, Patent Law, 6th edition 2021, para. 859.

3.11 Function-Oriented Interpretation

A skilled person will interpret a claim in such a way that an appropriate technical effect or function results. The skilled person will therefore look for the technical function of the claim. Function-oriented interpretation thus aligns with the way of thinking and working of the skilled person.[15]

However, caution is required with this variant of interpretation, as it is possible that an assumption is made about a function and the meaning of the claim is interpreted in a direction that does not correspond to the claim wording. It should be examined whether the result of the function-oriented interpretation is directly and unambiguously derivable for the skilled person from the originally filed application documents and the claim to be interpreted.[16] Preferably, function-oriented interpretation should be used affirmatively to confirm an already established understanding of a claim wording.

The skilled person would not assign technically nonsensical embodiments to the scope of protection of a claim. Function-oriented interpretation can be used in particular to exclude such technically nonsensical embodiments, as it asks about the function and can thus identify embodiments without function or with nonsensical functions.[17]

3.12 Interaction of Interpretation Variants

The interpretation variants are not applied alternatively. Instead, all methods are used wherever possible in order to achieve an appropriate overall interpretation of the respective patent claim. The interpretation variants are therefore used in combination. It should also be noted that, in principle, there is no hierarchy among the interpretation methods, whereas the significance of the individual interpretation methods may vary in each individual case.

Table 3.1 shows the four interpretation variants, whereby a literal interpretation should always be performed first, followed by a systematic interpretation based on the description and drawings. The teleological and functional interpretation variants are to be applied rather as supplements.

---

[15] Timmann, Section 3. Interpretation and Scope of Protection of Patents, Haedicke/Timmann, Handbook of Patent Law, 2nd edition 2020, para. 51.

[16] Benkard PatG/Scharen, 12th ed. 2023, PatG § 14 para. 71; Meier-Beck GRUR 2003, 905, 907.

[17] Timmann, Section 3. Interpretation and Scope of Protection of Patents, Haedicke/Timmann, Handbook of Patent Law, 2nd edition 2020, para. 52.

Table 3.1 Interpretation variants of a claim

Interpretation variants of a claim	Method	Significance
Literal/word-for-word interpretation	Interpretation by adding the literal meaning of the individual terms of the claim	Always to be applied
Systematic interpretation	Interpretation in the context of the description and drawings	Always to be applied
Teleological interpretation	What task is the invention intended to fulfill?	Only supportive
Function-oriented interpretation	What technical effect should the embodiments have?	Only supportive

3.13 Extension of the Scope of Protection through Equivalence

Despite the best efforts, a claim formulation may fail to describe particular future forms of use that are encompassed by the essence of the patented invention. In particular, functionally equivalent substitutes should not be excluded from the wording of the claim. It would be inappropriate to regard such equivalent forms of use as outside the scope of protection. For this reason, the scope of protection of a patent claim must be extended to cover equivalent embodiments.

However, this must not result in an unjustified subsequent extension of the scope of protection that cannot be derived directly and unambiguously from the wording of the claim itself. Therefore, any extension of the scope of protection through equivalence must be carried out with great caution. Strict limits must be observed, which are reflected in three criteria.

The equivalently extended scope of protection of a claim also includes those embodiments that the skilled person, based on the technical teaching formulated in the claim, considers to be equivalent, having the same effect, and obvious.

The scope of protection of a claim can be extended to forms of use that meet the following criteria:

- **Same effect:** The alternative form of use is characterized by substitute features that have the same effect.
- **Obviousness:** The substitute features are obvious to the skilled person. If replacing the original features requires inventive activity, equivalence cannot be assumed.
- **Equivalence:** The skilled person must regard the substitute features as equivalent to the original features.[18]

[18] BGH 12.3.2002, X ZR 168/00—Schneidmesser I; BGH 12.3.2002, X ZR 135/01—Schneidmesser II; BGH 12.3.2002, X ZB 12/00—Custodiol I; BGH 12.3.2002, X ZR 73/01, Custodiol II; BGH 12.3.2002, X ZR 43/01—Kunststoffrohrteil; BGH GRUR 1987, 279, Formstein; BGH GRUR 2015, 361 Kochgefäß; BGH GRUR 2016, 1031 Wärmetauscher.

The criterion of obviousness can be determined using standard methods, in particular the problem-solution approach in conjunction with the could-would test.[19]

Teleological and Functional Interpretation for Determining Equivalence
Teleological and functional interpretation can be used to identify equivalent substitute features. Functional interpretation, in particular, can be used to determine substitute features that are characterized by having the same effect as the original features. Teleological interpretation, which considers the purpose and intent of a feature, can be used to identify equivalent features.

Formstein Objection
However, the scope of protection cannot be extended to equivalent forms of use that demonstrably belong to the public domain prior art. This limitation of the scope of protection arises from the Formstein objection.[20] Prior art therefore constitutes an additional restriction on the scope of protection by equivalence.

3.14 Definitions

If the application contains definitions, these are decisive for the meaning of the corresponding terms. Generally accepted and customary meanings of terms are to be disregarded in this case. A patent specification can serve as its own lexicon or dictionary. An understanding that arises from the patent specification and differs from the customary technical language must therefore not be ignored.[21]

3.15 Interpretation of the Application Documents

The scope of protection of a patent is determined by the patent claims.[22] However, in order to determine the meaning of the claims, the description and the drawings must be taken into account, so that, if necessary, the application documents or the entire patent specification must be interpreted according to the understanding of the skilled person.[23]

[19] EPO, https://www.epo.org/de/legal/guidelines-pct/2025/g_vii_5.html, accessed on 18.8.2025; EPO, https://www.epo.org/de/legal/guidelines-pct/2025/g_vii_5_3.html, accessed on 18.8.2025.

[20] BGH GRUR 1987, 279, Formstein.

[21] BGH 2.3.1999, X ZR 85/96, GRUR 1999, 909—Spannschraube.

[22] § 14 sentence 1 Patent Act.

[23] § 14 sentence 2 Patent Act; BPatG 11.10.2016, 4 Ni 7/15—Bioreactor; BGH 2.3.1999, X ZR 85/96, GRUR 1999, 909—Spannschraube.

3.16 Non-Technical versus Technical Features

Non-technical features are generally disregarded. However, a different situation arises if the inventive step only results from the interaction of the technical and non-technical features. In this case, the non-technical features must be taken into account when interpreting the claims.

3.17 Difference Between Interpretation and Subsumption

A granted claim is to be regarded as a legal norm, which is why interpretation of the claim and subsumption are fundamentally necessary. Interpretation determines the meaning of the claims. In subsumption, it is then clarified whether a particular technical device or a technical manufacturing or application process falls under the interpreted meaning of the relevant claim.

Subsumption constitutes the "Reasoning" section of an opposition brief, a nullity action, or an infringement action, in which it is described that the features of the claim have been anticipated in a novelty-destroying manner by the prior art document or are not based on inventive step. In patent infringement proceedings, subsumption serves to justify that the infringer's product infringes the features of the claim.

Table 3.2 shows the essential differences between interpretation and subsumption.

Interpretation precedes subsumption. First, the meaning of the claim is determined in order to identify the embodiments covered by the claim. In the second step, subsumption follows, in which it is determined whether, in an opposition or nullity proceeding, the prior art challenges the validity of the patent. In the grant procedure, subsumption is used to assess patentability.

Table 3.2 Comparison of Interpretation and Subsumption

	Interpretation	Subsumption
Purpose	Determining the meaning of a claim	Does a product fulfill the features of a claim (patent infringement), or are the features of the claim contained in one or more documents (validity)?
Method	Literal, systematic, teleological, and function-oriented interpretation	Major premise (formulation of the interpreted claim), minor premise (element corresponds to the feature of the claim), and conclusion (the feature is disclosed)
Example	The feature of the claim describes a fastening element for a picture on a wall	**Major premise:** The claim describes in the feature a fastening element for a picture on a wall **Minor premise:** In document X, a nail is disclosed as a fastening element for a picture to be attached to a wall **Conclusion:** Document X discloses the feature

Table 3.3 Examples of Subsumption

Examples of Subsumption	Apparatus Claim	Method Claim
Major premise	Apparatus with features A, B, and C	Method with steps A, B, and C
Minor premise	The product XY fulfills features A, B, and C	The method XY comprises steps A, B, and C
Conclusion	The product XY falls within the scope of protection of the claim	The method XY falls within the scope of protection of the claim

Subsumption in patent law takes place in three steps, with the first step based on the result of interpretation, i.e., the understanding of the patent claim. In the next step, it is examined whether an element falls within the scope of protection of the interpreted claim, and the concluding sentence summarizes the result.

Table 3.3 shows the different formulations of subsumption for a product claim/apparatus claim and a method claim.

4 Amendment of Claims

This chapter describes the fundamental framework conditions within which amendments to the claims are possible. These possibilities may be further restricted, particularly after the grant of the patent, by additional legal provisions.

4.1 Impermissible Extension

An impermissible extension of the subject matter of a patent application occurs if an amendment to the application cannot be derived from the originally filed application documents. Amendments that go beyond the disclosure of the originally filed application documents are not permitted.[1] No rights can be derived from an impermissibly extended subject matter.[2] Rather, an impermissible extension calls into question the legal validity of the patent.

A patent can only be granted for a technical teaching that has not been impermissibly extended. If the applicant does not remedy an impermissible extension, the patent application must be rejected by the Patent Office.[3]

A technical teaching can be derived from the application documents and therefore does not constitute an impermissible extension if it is "directly and unambiguously"

[1] Section 38 sentence 1 of the Patent Act.

[2] Section 38 sentence 2 of the Patent Act.

[3] Section 48 sentence 1 in conjunction with Section 45 paragraph 1 sentence 1 of the Patent Act.

T. H. Meitinger, *Drafting, Interpretation, and Amendment of Patent Claims*,
https://doi.org/10.1007/978-3-662-73622-7_4

disclosed in the originally filed application documents. The perspective of the skilled person is to be applied here.[4]

An impermissible extension can already arise from picking out a single feature from a combination of features and incorporating it into a claim. This is especially the case if the application documents do not disclose that this feature can also be used individually, and not only in the described combination of features. If the application documents additionally describe that this feature is inseparably linked to the combination of features, then using the individual feature in another claim combination constitutes an impermissible extension.[5] In this case, a new subject matter, an aliud, is created that was not described in the originally filed application documents.[6]

4.2 Extension of Scope of Protection

An extension of the scope of protection is to be considered independently of an impermissible extension. An extension of the scope of protection arises if the scope of protection of a granted patent is subsequently broadened or if embodiments are subsequently claimed that were not previously protected by the patent. These newly added embodiments may result from an impermissible extension. However, it is also possible that the new embodiments were contained in the originally filed application documents.

The prohibition of extension of the scope of protection is intended to protect the public from unexpected claims for injunctive relief. Once a third party has informed themselves about the scope of protection of a patent, they should not suddenly be confronted with additional patented embodiments. An extension of the scope of protection constitutes grounds for revocation in invalidity proceedings.[7]

---

[4] Federal Court of Justice (BGH), judgment of July 19, 2016—X ZR 36/14; BGH, judgment of December 16, 2008—X ZR 89/07, Olanzapine; BGH, judgment of July 8, 2010—Xa ZR 124/07, GRUR 2010, 910—Tamper-proof document; BGH, judgment of June 9, 2015—X ZR 51/13, GRUR 2015, 976 para. 45—Injection valve; EPO Enlarged Board of Appeal, Official Journal 2001, 413 = GRUR Int. 2002, 80; EPO GRUR Int. 2008, 511—Traction sheave elevator/KONE.

[5] BGH, decision of September 11, 2001—X ZB 18/00, GRUR 2002, 49—Torque transmission device; BGH, judgment of February 17, 2015—X ZR 161/12, BGHZ 204, 199 para. 31—Wound treatment device; BGH, judgment of June 21, 2016—X ZR 41/14, GRUR 2016, 1038 para. 48—Vehicle window II; BGH, judgment of September 26, 2023—X ZR 76/21, GRUR 2024, 42 para. 42—Color and brightness adjustment.

[6] BGH, judgment of October 16, 2007—X ZR 226/02—Sammelhefter II; BGH, BGHZ 66, 17, 29—Alkylendiamine I; BGH, BGHZ 110, 123, 125—Splice chamber.

[7] Section 22 paragraph 1, second alternative of the Patent Act.

Table 4.1 Comparison of impermissible extension and extension of scope of protection

	Meaning	German Patent Act	European Patent Convention (EPC)	Risk in German proceedings	Risk in European proceedings
Impermissible extension	An amendment cannot be derived from the originally filed application documents.	Section 21 paragraph 1 no. 4 Patent Act	Article 123 paragraph 2 EPC	Revocation in opposition or invalidity proceedings	Revocation in opposition proceedings
Extension of scope of protection	An amendment to the set of claims constitutes an extension or a change to an already granted scope of protection.	Section 22 paragraph 1, second alternative Patent Act	Article 123 paragraph 3 EPC	Disclaimer may be required	Inescapable trap

Table 4.1 describes the differences between impermissible extension and extension of the scope of protection.

4.3 Features from Different Embodiments

The combination of features from different embodiments constitutes an impermissible extension, even if the individual embodiments are contained in the originally filed application documents. This remains the case even if the combination of features is obvious, because what is not considered impermissibly extended is not what is obvious, but only what is "directly and unambiguously" disclosed.[8]

Furthermore, an amendment is impermissible if it results from the incorporation of a feature from the description and this constitutes a creative achievement.[9] Even an amendment that does not constitute a creative achievement and therefore appears obvious is not permissible if it does not result directly and unambiguously from the original application documents.[10]

[8] BGH July 8, 2010, Xa ZR 124/07, GRUR 2010, 910—Tamper-proof document.

[9] BGH January 23, 1990, X ZB 9/89—Splice chamber.

[10] BGH April 9, 2013, X ZR 130/11, GRUR 2013, 809—Encryption method.

4.4 Disclosure in Drawings

Drawings are considered part of the disclosure of a patent application in the same way, so the inclusion of features from the drawings is generally permissible. Therefore, if a feature is disclosed exclusively in the drawings and not otherwise in the application documents, it can still be validly incorporated into a claim. However, the corresponding feature must be clearly derivable from the drawings.[11]

Drawings do not have a subordinate significance. Drawings are not only used to illustrate features described in the application text. Rather, the drawings can even represent the central and original disclosure of the inventive teaching.[12]

However, it can be difficult to derive exact dimensions or size ratios from drawings.[13] In particular, if no precise dimensions are indicated, only qualitative features can be derived from a drawing.[14]

Disclosure by a drawing is especially unproblematic if it can also be assumed that the skilled person would have "read" the feature and the feature is not assigned a role essential to the invention.[15]

4.5 Obvious Amendment

Amendments to the application documents are not permissible merely because they are obvious. Features can only be incorporated into a claim without resulting in an inadmissible extension if it is evident that they belong to the invention. Features described in relation to another embodiment cannot be added to a claim. Furthermore, it is difficult to argue for the inclusion of features in a claim that are described in the application text as belonging to the prior art and as disadvantageous.[16]

Figure 4.1 illustrates that the incorporation of a feature from the description into a claim is possible if there are no obvious reasons against it. Of course, it is also permissible to incorporate features from dependent claims. In addition, features can be inserted that the skilled person would directly and unambiguously derive from the application text, even if they are not expressly disclosed.

[11] Federal Court of Justice, 30.1.2007, X ZR 156/02, GRUR 2007, 578, 580—Backwashable filter candle.

[12] Federal Court of Justice, 18.2.2010, Xa ZR 52/08, GRUR 2010, 599—Molded part.

[13] Federal Court of Justice, 16.10.2012, X ZB 10/11—Plug connection.

[14] Higher Regional Court of Vienna, 31.3.2016, 34 R 159/15z—Frequency catalyst.

[15] Federal Court of Justice, 24.1.2012, X ZR 88/09, GRUR 2012, 475—Electron beam therapy system.

[16] Federal Court of Justice, 23.1.1990, X ZB 9/89—Splice chamber.

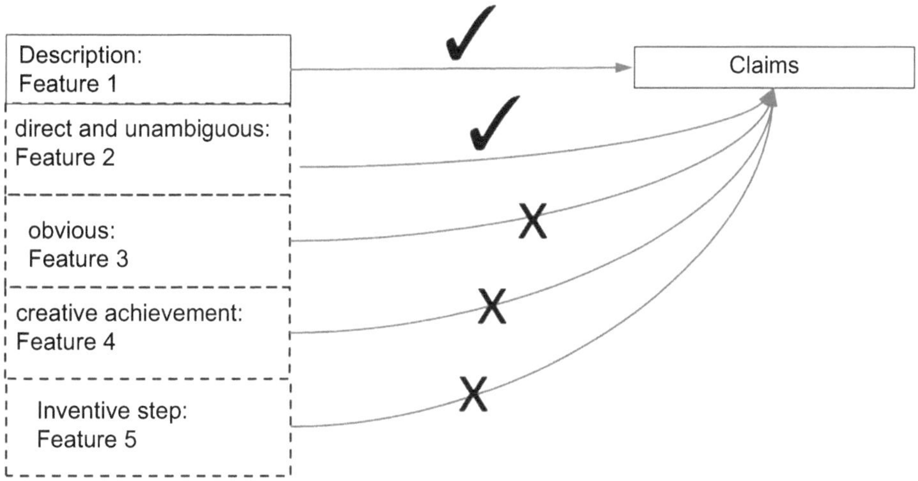

Fig. 4.1 Amendment of a set of claims

Features that can only be derived from the application documents through a creative achievement or inventive step can in no case be used to amend the set of claims. Such an approach would definitely constitute an inadmissible extension.

4.6 Amendment of Wording

An amendment of the wording of the claims is permissible as long as the meaning of the amended claims is covered by the disclosure of the originally filed application documents. Therefore, it is generally unproblematic to use synonyms. If a feature is described in detail in the description so that the skilled person can directly and unambiguously derive the feature, the appropriate technical term for this feature can be used, even if the term itself does not appear in the application documents.[17]

For example, if a component is described in detail in the application documents and only in the grant procedure are individual parts of the component designated in a manner that the skilled person would have used in at least a similar way, this addition by incorporating the designations does not constitute an inadmissible extension.[18]

[17] Federal Court of Justice, 29.4.2014, X ZR 19/11, GRUR 2014, 970—Stent.

[18] Federal Court of Justice, 21.9.1993, X ZR 50/91, Mitteilungen der deutschen Patentanwälte 1996, 204, 206—Track section.

However, replacing a term must not result in a change of meaning (aliud) or an expansion of meaning. Therefore, changes should only be made when necessary and not without good reason, in order to achieve a supposed "clarification" of the claimed scope of protection. Amending the wording always carries the risk of inadmissible extension or extension of the scope of protection. The risk to the legal validity of the patent can therefore never be ruled out.

4.7 Correction of Errors

Correction of errors is generally possible at any time, bearing in mind that application documents do not have to be error-free and that errors can also creep in during the grant procedure.[19] Obvious errors are therefore tolerated at any stage of the proceedings, even without correction.

4.8 Correction of Translations

Patent applications can be filed at the German Patent and Trademark Office in any language. However, a German translation must be submitted within a period of three months.[20] If the translation is incorrect, a corrected version can be submitted to the Patent Office at any time, and there is no risk to the application or the filing date of the application.[21]

The same essentially applies to application documents filed at the European Patent Office.[22]

[19] Meier-Beck, The case law of the Federal Court of Justice on patent and utility model law in 2011, GRUR 2011, 1177, 1181.

[20] Section 35a paragraph 1 sentence 1 Patent Act.

[21] Federal Court of Justice, judgment of 18 July 2011—X ZB 10/10—Determination of polishing point.

[22] Article 14 paragraph 2 European Patent Convention (EPC); EPO, https://www.epo.org/de/service-support/faq/anmeldung-eines-patents/praxis-tipps/ich-habe-die-auskunft-erhalten-dass, accessed on 6.8.2025.

5 Amendment During the Grant Procedure

The amendment of claims is generally permissible within the disclosure of the originally filed application documents. However, if features are subsequently introduced, this results in an unjustified advantage for the applicant, which constitutes an inadmissible extension and is not permitted.

5.1 Features from the Description

The incorporation of features from dependent claims into the referenced independent claim is unproblematic. When features from the description are incorporated into the claims, certain requirements must be observed.[1]

If features are explained in the description to clarify or illustrate the subject matter of a claim, the incorporation of these features into the respective claim is permissible.[2] However, it must always be clear and unambiguous to the skilled person that the resulting combination of features was contained in the originally filed application documents.[3]

[1] Federal Court of Justice (BGH), judgment of November 25, 2014—X ZR 119/09—Abrasive Product; with reference to BGH, judgment of February 11, 2014—X ZR 107/12, BGHZ 200, 63 = GRUR 2014, 542 para. 24—Communication Channel.

[2] BGH, judgment of November 25, 2014—X ZR 119/09—Abrasive Product; with reference to BGH, decision of January 23, 1990—X ZB 9/89, BGHZ 110, 123, 126—Splice Chamber; BGH, judgment of August 30, 2011—X ZR 12/10, para. 30.

[3] BGH, judgment of November 25, 2014—X ZR 119/09—Abrasive Product; with reference to BGH, decision of September 11, 2001—X ZB 18/00, GRUR 2002, 49—Torque Transmission Device.

T. H. Meitinger, *Drafting, Interpretation, and Amendment of Patent Claims*,
https://doi.org/10.1007/978-3-662-73622-7_5

5.2 Features of Different Embodiments

An application may describe only a single invention, whereby the included embodiments must represent different realizations of the common inventive concept.[4] If this commonality is not present, there is a lack of unity. Unity exists only if the different embodiments of a patent application agree in those features that constitute the essential distinction from the prior art.[5]

A description of a patent application contains different embodiments that deviate from each other and represent further developments of the invention in different directions. These embodiments may realize the same common inventive concept and yet differ so significantly from each other that their technical teachings cannot readily be combined. Therefore, only those embodiments of a patent application whose combination is clearly and unambiguously derivable from the application documents may be combined. Conversely, if a combination is expressly excluded in the patent application, this combination cannot be described in a claim as the claimed subject matter. However, if it is obvious to a skilled person that features of different embodiments can be combined, such a combination is not precluded by the criterion of inadmissible extension. This also applies if the new claim formulation points in a different direction and serves the realization of another technical effect.[6]

However, a combination of features from different embodiments may be rejected if the relevant features exhibit special effects and the application documents do not disclose how these effects are to be combined to achieve the desired inventive effect.[7]

A combination of very different embodiments or a combination of features from different embodiments is only unrestrictedly permissible if the application documents provide an indication of such a combination. It must be clear and unambiguous from the application documents that the combination is encompassed by the application documents.[8]

5.3 Generalization

The grant procedure serves to achieve the broadest possible scope of protection for the applicant in light of the prior art. In this context, the question of abstraction regularly arises, in which embodiments described in the application are taken as a starting point

[4] Section 34 paragraph 5 Patent Act; Article 82 EPC.

[5] Rule 44 paragraph 1 EPC.

[6] BGH January 30, 2018, X ZR 27/16—Waterproof Leather Shoe.

[7] BGH May 14, 2009, Xa ZR 148/05—Heater.

[8] EPO, https://www.epo.org/de/legal/case-law/2022/clr_ii_e_1_6_2_c.html, accessed on 1.7.2025; EPO T 686/99; T 714/08.

and individual features are to be omitted. Such an approach is permissible if the corresponding claim formulation is directly and unambiguously derivable from the original application documents. The assessment method of "direct and unambiguous disclosure from the perspective of the skilled person" is referred to by the European Patent Office as the "gold standard."

The European Patent Office is currently rather skeptical to dismissive regarding the deletion of features from filed claims or the addition of features from other embodiments. Furthermore, if a feature is described in the application as necessary for the realization of the invention, the European Patent Office will not accept a claim without this feature.[9]

In German proceedings, a situation more favorable to the applicant can be assumed. The Federal Court of Justice has pointed out the preliminary nature of claim formulations in the grant procedure and has confirmed the legitimate interest of the applicant in obtaining the broadest possible scope of protection, including through generalizations.[10] If a feature is not decisive for solving the technical problem, it can more readily be generalized based on individual embodiments with specific details. The Federal Court of Justice even considers it acceptable for a feature to be subsequently deleted if it is obviously not necessary for the skilled person to realize the invention. In the decision "Torque Transmission Device", the Federal Patent Court followed this line and accepted that features not required for the invention from the skilled person's perspective were deleted.[11] A generalization may even be permissible based on only a single embodiment. This also applies if other technical implementations are conceivable.[12]

In the "Splice Chamber" decision of the Federal Court of Justice, it was established that the applicant has considerable freedom to use individual features from exemplary embodiments to distinguish from the prior art. The Federal Court of Justice therefore grants the applicant the right to limit their patent in any desired form.[13]

In the decision "Fat Suction Device"[14] an exemplary embodiment with a component was described in detail, but its function was not mentioned. The Federal Court of Justice accepted the abstraction of the feature to its function, since the skilled person can reproducibly achieve the function. It is therefore not necessary for the function to be explicitly described in the application documents.

---

[9] EPO T 1515/11.

[10] BGH 11.2.2014, X ZR 107/12—Kommunikationskanal.

[11] BPatG 19.6.2007, 6 W (pat) 40/04—Torque Transmission Device.

[12] BGH 16.10.2007, X ZR 226/02—Sammelhefter II.

[13] BGH 23.1.1990, X ZB 9/89—Splice Chamber.

[14] BGH 20.12.2016, X ZR 84/14—Fat Suction Device.

In particular, the applicant should be allowed a more concise claim formulation by deleting features if the meaning of the deleted feature is already provided by the remaining features.[15]

However, the deletion of a feature or the replacement of a feature with a more general feature is only permissible if it can be inferred from the description that the deleted feature is not essential to the invention. The description must show that the invention can also be realized without the relevant feature.[16] A generalization is particularly excluded if it does not "readily" follow for the skilled person.[17]

5.4 Amendment Due to Lack of Clarity

If, during the grant procedure, the examiner at the Patent Office determines a lack of clarity, the applicant is requested to reword the claims.[18] However, the applicant should be cautious and not choose a wording that results in an inadmissible extension. The applicant should, if possible, find a claim wording that can be taken verbatim from the description.

When weighing a lack of clarity in the claims against the risk of an inadmissible extension, the applicant should, in case of doubt, be allowed to retain their original wording. In this case, the ground for revocation due to inadmissible extension is safely avoided. It should also be noted that lack of clarity is not a ground for opposition or revocation. Therefore, if there is a choice between the risk of inadmissible extension and alleged lack of clarity, the applicant should preferably avoid the risk of inadmissible extension. However, if the existing lack of clarity is unacceptable and sufficient clarity cannot be achieved without inadmissible extension, the Patent Office must reject the patent application.

Clarity is required because otherwise an exact determination of the scope of protection of the claims is not possible.[19] If there are ambiguities, these always work to the detriment of the applicant, so that in infringement proceedings a narrow and in opposition and nullity proceedings a broad interpretation of the claims is applied, if, due to lack of clarity after interpretation by the skilled person, both a narrow and a broad understanding are possible.

[15] BPatG 5.4.2005, 21 W (pat) 303/03, Mitteilungen der deutschen Patentanwälte 2006, 28.

[16] BGH 22.12.2009, X ZR 27,28/06—Hubgliedertor I, II.

[17] Meier-Beck, The case law of the BGH on patent and utility model law in 2012, GRUR 2013, 1177, 1180; BGH 14.8.2012, X ZR 3/10—UV-sensitive printing plate.

[18] § 34 paragraph 4 Patent Act.

[19] Einsele, Drafting patent claims—clear or not clear, Mitteilungen der deutschen Patentanwälte 2014, 249, 254.

5.5 Amendment Not Possible Despite Disclosure

It is questionable whether every disclosed embodiment can be used to amend claims. This may be denied even though the embodiment is comprehensively described and can be carried out by a skilled person. If embodiments are described as disadvantageous or as belonging to the prior art, for example, it cannot be assumed that the inventor considered these as part of the claimed invention. Accordingly, the applicant cannot make such designated embodiments part of their request for protection. Moreover, an infringement action based on these embodiments is ruled out from the outset.[20]

Phrases such as "essential element of the invention" or "belonging to the invention" should be used with great caution in the application documents, as it will be difficult in the grant procedure to omit such designated features from the independent claims in order to obtain a broader scope of protection. The examiner at the Patent Office will attribute an essential inventive role to features that are assigned significant importance in the application documents and will require that these be mentioned in the main claim.

The inventor may be mistaken in their invention in that they did not understand why their invention achieves the desired technical effect. The inventor may allow themselves a lack of understanding of the invention without endangering the grant of a patent. Nevertheless, the intention expressed in the application that a feature belongs to the invention is taken into account, so that omission of this feature is not permissible.

The same applies to embodiments that are included in the description but are expressly described in the application documents as not belonging to the claimed invention. An extension of the scope of protection based on such embodiments can hardly be justified any longer.

5.6 Amendment Possible Despite Impermissible Extension

Features without technical content are not taken into account when determining patentability.[21] Such a feature exists, for example, when it is specified what color a signal lamp has. If, in this case, it is specified that a green or blue light appears instead of a red or yellow one, this constitutes a non-technical feature. This feature may have been added later and not disclosed in the originally filed application documents, thus constituting an impermissible extension. Nevertheless, such a feature can be tolerated and left in a claim, as it has no impact on the technical substance of the subject matter of the claim. A feature that has no effect on novelty or technical function can remain untouched despite

[20] Ackermann, Selection decisions in patent grant and validity proceedings, Mitteilungen 2017, 195.

[21] EPO, https://www.epo.org/de/legal/guidelines-epc/2024/g_vii_5_4.html, accessed on 7/6/2025.

being an impermissible extension, since the impermissible extension exists but is to be regarded as insignificant.[22]

Additionally, a feature that does not provide the applicant with any advantage can be tolerated in a claim, even if it constitutes an impermissible extension. It is not necessary to declare a patent invalid due to a feature that does not constitute an undue reward for the applicant. However, this only applies if the additional feature does not introduce a new technical teaching into the claim. A deterioration of the technical teaching is also not permissible, as it affects the technical substance of the invention and leads to a distinction from the prior art. An impermissible extension by way of deterioration therefore cannot be accepted and may lead to revocation in opposition or nullity proceedings.

5.7 Change of Category in Grant Proceedings

In grant proceedings, a change of category is permissible as long as it does not result in an impermissible extension. However, the applicant should in any case include both a product claim and a process claim in their set of claims. It is advisable to always describe the inventive concept both as an apparatus claim and as a method claim in order to claim the broadest possible scope of protection. This also provides the opportunity to use "hybrid formulations" to broaden the scope of protection. If the individual steps by which the inventive device is manufactured are described in the specification, for example, it becomes possible to include a "product-by-process" claim in the set of claims.

[22] BPatG 11.10.2016, 4 Ni 7/15—Bioreactor.

6 Amendment After Grant

While in the grant phase before the patent office, an amendment of the application documents is possible at any time and is also envisaged in interplay with the office actions of the patent office, this does not apply to the period after the patent has been granted. In the pre-grant phase, the only requirement is to avoid any inadmissible extension. After grant, the applicant is in particular no longer permitted to enlarge the scope of protection. The prohibition of extension of the scope of protection applies. At most, a reduction of the scope of protection is possible, for example to respond to newly discovered prior art.

The inclusion of features in the independent claims to reduce the claimed scope of protection is granted to the patent proprietor, especially in contentious proceedings. The applicant may limit the scope of protection of their patent in opposition or nullity proceedings. Outside of contentious proceedings, the applicant is also allowed to reduce the scope of protection, for example as a precaution to avoid contentious proceedings.

Furthermore, the patent proprietor is permitted to renounce parts of their patent that are no longer of importance to them. This can bring a patent into a form that facilitates enforcement. In some cases, an amendment of the claims may make an infringement more evident, thereby accelerating the granting of a preliminary injunction or judicial patent enforcement.

An amendment of the claims is only possible if it does not result in an inadmissible extension, i.e., the disclosure of the patent specification does not go beyond what was originally disclosed.

While an amendment of the patent claims during the grant procedure "only" carries the risk of resulting in an inadmissible extension, amendments after grant can even endanger the validity of the patent. An inadmissible extension in the grant procedure can be remedied by the patent proprietor in opposition or nullity proceedings. A change in the scope of protection with a feature that was not contained in the originally filed documents or that is not disclosed in combination with the remaining claim features in the

T. H. Meitinger, *Drafting, Interpretation, and Amendment of Patent Claims*,
https://doi.org/10.1007/978-3-662-73622-7_6

original application documents can result in the revocation or nullification of the entire patent without any possibility of remedy. This applies in particular to the European procedure, where the so-called "inescapable trap" threatens.

6.1 Inadmissible Extension of the Patent

The technical teaching of the patent may not be extended in such a way that it is no longer directly and unambiguously derivable for the skilled person from the originally filed application documents.

6.2 Extension of the Scope of Protection

Extension of the scope of protection means that a set of claims already granted is amended in such a way that the new set of claims has a broader or different scope of protection. In this case, the legal protection interests of the public are violated. The public must be able to rely on the fact that a granted scope of protection will not be enlarged.

A particular embodiment of a patent is included in the scope of protection if the embodiment results from the wording of the claims. It is not sufficient if the embodiment can be derived from the entire application documents. The claims therefore determine the scope of protection of a patent. However, the description and the drawings must not be ignored, as they serve to aid the understanding of the claim wording and must be taken into account. This applies even if the claim wording appears clear and unambiguous.[1] However, if the disclosure of the description contradicts the claim wording, only the claim wording is to be considered.

An extension of the scope of protection cannot result from dependent claims, since the scope of protection of a patent is determined exclusively by the independent and not by the dependent claims.

6.3 Amendments Under German Law

The patent proprietor may revoke their patent at any time or make a limitation of the patent claims. The limitation applies retroactively (ex tunc).[2]

Additionally, the patent proprietor may defend their patent with an amendment of the claims in opposition or nullity proceedings, provided that this does not constitute an inadmissible extension or an extension of the scope of protection.

[1] Section 14 sentence 2 Patent Act.

[2] Section 64 paragraph 1 Patent Act.

6.4 Amendments Under European Law

Under European law, there are three possibilities for amending the claims. According to Article 105a paragraph 1 sentence 1 EPC, the applicant may completely revoke their patent or amend the patent claims in such a way that the patent is limited.

Furthermore, in opposition proceedings, the applicant may file a request for maintenance of their patent in a limited form and attach amended claims to their request. The request may be filed as an auxiliary request.[3]

There is also the possibility to correct linguistic errors, clerical errors, and obvious mistakes in decisions of the European Patent Office.[4] However, the significance of this possibility for amendment is very limited, as no substantive corrections are permitted.[5]

Article 123 EPC deals with the possibilities for amending a patent application or a patent before the European Patent Office. The applicant must be given at least one opportunity to amend their patent application.[6] However, a new version of the patent application or the patent must be supported by the originally filed application documents.[7] Furthermore, a scope of protection once granted may not be subsequently extended by amendments.[8]

Article 123 paragraph 2 EPC prevents completely new features from being introduced into the patent specification or patent application. In addition, combinations of disclosed features that were not originally combined in this way are excluded. Unexpected abstractions are also not permitted. Article 123 paragraph 2 therefore prevents inventions from being claimed that did not exist on the filing date.

So-called intermediate generalizations, or more generally, generalizations that cannot be derived from the original application documents, are generally refused by the European Patent Office on the basis of Article 123 paragraph 2 EPC.

An intermediate generalization is the mixing of two or more embodiments, resulting in a new embodiment that has not previously been described or claimed in this combination in the application documents. An intermediate generalization also arises if a feature is removed from a described embodiment. The removal of a single feature from an embodiment is only permissible if the feature is not in a structural or functional relationship with other features of the embodiment.[9]

---

[3] Article 101 paragraph 3 EPC.

[4] Rule 140 EPC.

[5] EPO G 2/10, 30.8.2011, OJ EPO 2012, 367—Disclaimer/Scripps.

[6] Article 123 paragraph 1 sentence 2 EPC.

[7] Article 123 paragraph 2 EPC.

[8] Article 123 paragraph 3 EPC.

[9] EPO, https://www.epo.org/de/legal/guidelines-epc/2024/h_v_3_2_1.html, accessed on 3.7.2025.

6.5 Inescapable Trap in Europe Versus Disclaimer in Germany

In German opposition proceedings, a patent is revoked if the disclosure of the patent goes beyond what was originally contained in the application documents.[10] If, in invalidity proceedings, it is determined that a feature of a claim was not originally disclosed, it must be removed pursuant to Section 21(1) No. 4 Patent Act in order to eliminate the inadmissible extension. However, if removing the feature results in an extension of the scope of protection of the patent, the patent may be declared invalid on the basis of Section 22(1) second alternative Patent Act.

The Federal Court of Justice had to assess such a "tricky" situation in the decision "Angle Measuring Device", weighing the public interest in legal certainty—embodied in the principle that the scope of protection of a granted patent may not be extended—against the patent proprietor's interest in maintaining their patent. The Federal Court of Justice considered both requirements to be met by reintroducing the originally undisclosed feature into the claim, but not taking it into account when assessing legal validity.[11] Such a solution to the conflict between Section 21(1) No. 4 Patent Act and Section 22(1) second alternative Patent Act is only permissible if it constitutes an admissible limitation. However, if the originally undisclosed feature results in a different subject matter of the patent (aliud), the patent must be declared invalid.[12]

Such a dilemma can also arise in European proceedings due to Articles 123(2) and 123(3) EPC, whereby a limiting but inadmissibly extended feature must be removed under Article 123(2) EPC, but cannot be omitted under Article 123(3) EPC in order to prevent an extension of the scope of protection. The Enlarged Board of Appeal of the European Patent Office considers both Article 123(2) EPC and Article 123(3) EPC to be equally binding, so that one provision cannot be subordinated to the other. Both legal norms must be fulfilled. If this is not possible, the patent in question must be revoked.[13] In European proceedings, this can therefore result in a so-called "inescapable trap."

Figures 6.1 and 6.2 illustrate the inescapable trap of Article 123(2) and (3) of the European Patent Convention. In Fig. 6.1, the thickly outlined polygon represents the granted scope of protection of the main claim. The prior art documents 1 and 2 and feature X restrict the scope of protection to the polygon with the thick border lines. However, feature X was not included in the originally filed application documents, so

---

[10] Section 21(1) No. 4 Patent Act.

[11] Federal Court of Justice, decision of 21 October 2010—Xa ZB 14/09—Angle Measuring Device.

[12] Federal Court of Justice, decision of 21 October 2010—Xa ZB 14/09—Angle Measuring Device; with reference to Federal Court of Justice GRUR 2001, 140, 141—Time Telegram.

[13] EPO Enlarged Board of Appeal, OJ EPO 1994, 541 = GRUR Int. 1994, 842 para. 13—Limiting Feature/Advanced Semiconductor Products.

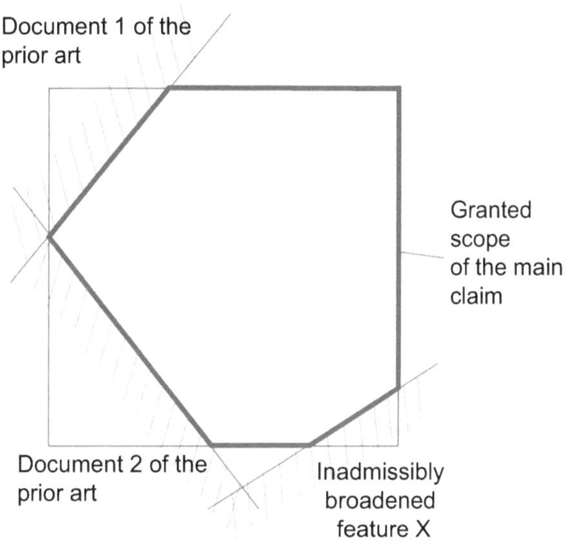

Fig. 6.1 Inadmissible Extension

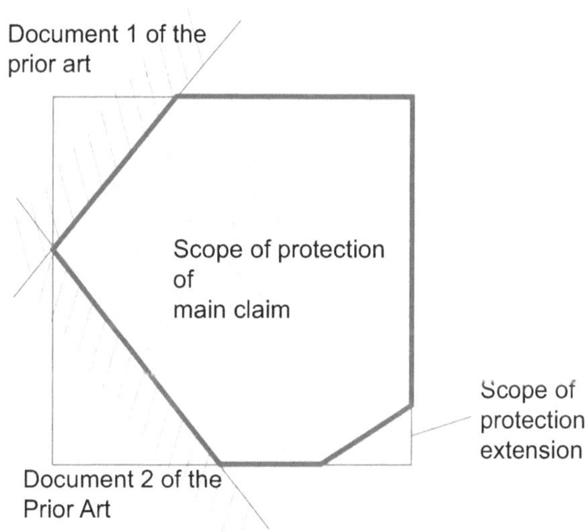

Fig. 6.2 Extension of Scope of Protection

according to Article 123(2) EPC, feature X must be removed from the main claim. Figure 6.2 shows the situation without feature X.

Figure 6.2 shows the diagonally hatched scope of protection of the main claim, where, due to the omission of feature X, the lower left corner outside the thickly outlined polygon has been added to the scope of protection, resulting in an extension of the scope of protection. According to Article 123(3) EPC, an extension of the scope of protection is not permitted, so feature X must be reintroduced into the main claim. Such an

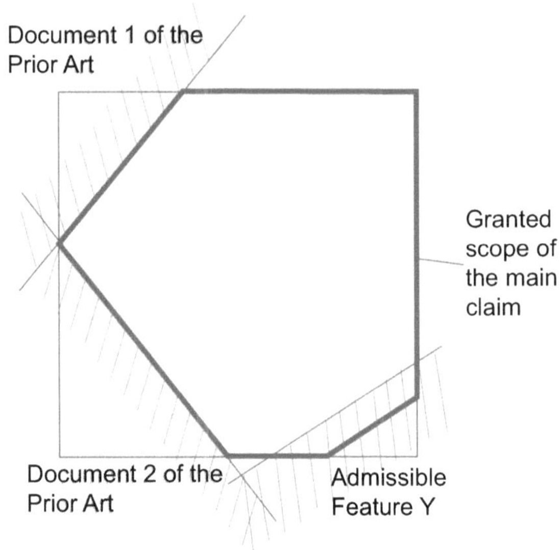

Fig. 6.3 Rescue from the "inescapable trap"

"inadmissible limitation" (by feature X) leads to an inescapable trap and the patent must be revoked in its entirety.[14]

The only way out of the inescapable trap is a feature Y that was included in the originally filed application documents and with which at least the area representing an extension of the scope of protection can be completely excluded from the scope of the main claim. Figure 6.3 shows such a feature Y, which can rescue the patent from the "inescapable trap."

It is clear that in most cases the patent specification does not contain a suitable feature to escape the inescapable trap. As a rule, an inescapable trap leads to the revocation of the European patent. In such cases, consideration should be given to a utility model branching in order to obtain at least utility model protection for Germany.[15]

6.6 Waiver Under German Law

The patent holder may, at any time, waive their patent or parts thereof pursuant to Section 20(1) No. 1 of the Patent Act. The waiver becomes effective on the day the declaration is made to the German Patent and Trademark Office. The waiver therefore applies prospectively and not retroactively to the filing date (ex nunc).

[14] EPO G 1/93, 2.2.1994, https://www.epo.org/de/boards-of-appeal/decisions/g930001dp1, accessed on 21.8.2025.

[15] Section 5(1) sentence 1 Utility Model Act; Meitinger, The Utility Model Branching: Rescue in German and European Opposition Proceedings? Mitt. 2021, July/August, p. 305.

With a waiver, the scope of protection sought can be reduced or relinquished entirely, thereby preempting an attack through opposition or nullity proceedings.

If the waiver is declared during nullity proceedings and the declaration of waiver is made within the one-month objection period under Section 82(1) of the Patent Act, this is to be regarded as an immediate acknowledgment by the patent holder. If, moreover, the defendant was not requested by the plaintiff to surrender the patent prior to the filing of the nullity action and otherwise gave no cause for the action, the plaintiff is to bear the costs of the nullity proceedings.

However, if the waiver by the patent holder occurs only later in the nullity proceedings, this is interpreted as an admission of defeat.[16] The cost consequences under the Code of Civil Procedure, which apply to the litigation costs of the nullity proceedings, stipulate that in this case the defendant is to bear the costs of the proceedings to the extent of the waiver.[17]

In German opposition proceedings, the Code of Civil Procedure does not apply. Instead, the opposition division may, at its equitable discretion, decide on the allocation of costs to a party.[18] The opposition division may impose costs on the patent holder in the event of a waiver, but this is not mandatory.[19]

Below are some examples of waiver declarations. For example, with the following set of claims:

1. Device for…
2. Device according to claim 1, wherein…
3. Device according to claim 1 or 2, wherein…
4. Device according to any of the preceding claims…

If claim 1 is waived, the situation depicted in Fig. 6.4 arises. In this case, a lack of unity in the set of claims may occur, since claims 2, 3, and 4 describe not only subject matter with common features (claim 1 $\wedge$ claim 2 $\wedge$ claim 3 $\wedge$ claim 4), but also subject matter that does not share common features.

Figure 6.5 illustrates a situation after the main claim has been removed, in which claims 2, 3, and 4 have no common features at all and, therefore, a lack of unity exists for each embodiment of the claim set. Lack of unity contradicts the principle that a patent may claim only one invention.

---

[16] Federal Patent Court, 5.5.2014–3 Ni 26/12 (EP); Federal Court of Justice, GRUR 61, 278, 279—Lamp housing.

[17] Section 84(2) sentence 2, first half-sentence of the Patent Act in conjunction with Section 91 ZPO.

[18] Section 62(1) sentence 1 of the Patent Act.

[19] Section 62(1) sentence 2, second half-sentence of the Patent Act.

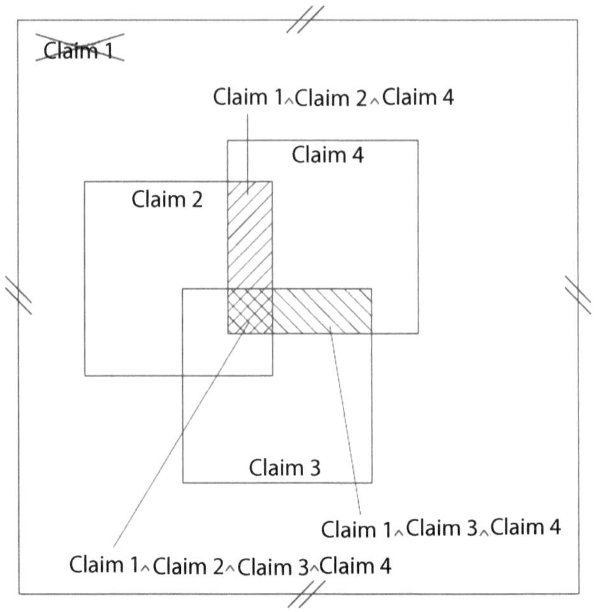

Fig. 6.4 Waiver of main claim with an intersection of the scopes of protection

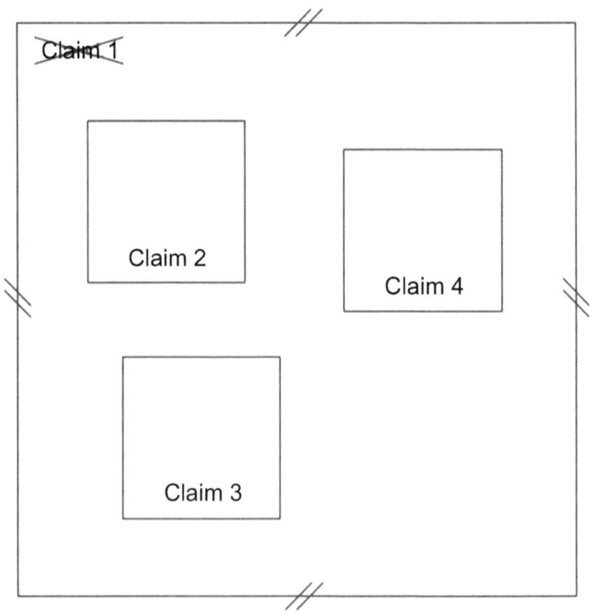

Fig. 6.5 Waiver of main claim without intersection of the scopes of protection

With this set of claims, a particular lack of unity arises for the dependent claims if the applicant waives the main claim or if it is revoked in contentious proceedings:

1. Device for…
2. Device according to claim 1, wherein…
3. Device according to claim 1, wherein…
4. Device according to claim 1, wherein …

It should be noted that lack of unity is neither a ground for opposition nor for revocation. Therefore, a set of claims cannot be challenged in opposition or nullity proceedings on the basis of lack of unity.

Only with a set of claims structured as shown in Fig. 6.6 does the deletion of the main claim (claim 1) not result in a lack of unity.

1. Device for…
2. Device according to claim 1, wherein…
3. Device according to claim 2, wherein…
4. Device according to any of claims 2 or 3, wherein…
5. Device according to any of claims 2, 3, or 4, wherein …

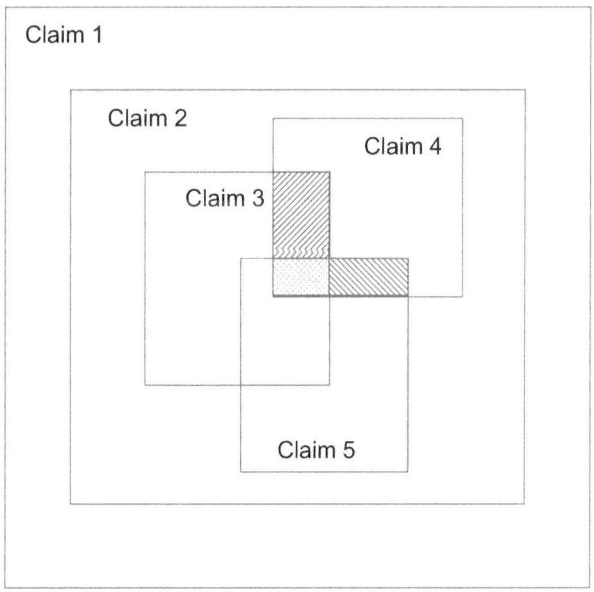

Fig. 6.6 Waiver of main claim without lack of unity

There is no waiver in European proceedings.[20] Instead, the applicant may limit their patent or request revocation.[21]

6.7 Limitation and Revocation

The patent holder may file a request for limitation or revocation in both the German and European proceedings. Through limitation, the scope of protection can be reduced by amending the patent claims. Revocation results in the complete cancellation of the patent.[22] The limitation or revocation has retroactive effect (ex tunc).[23]

In opposition or nullity proceedings, a patent holder may declare that they will no longer defend their patent to a certain extent or that they are surrendering the patent entirely. This constitutes a de facto revocation or limitation, thereby restricting the subject matter of the dispute. To the extent that the patent is surrendered, the proceedings are concluded without further substantive examination, and the patent is to be declared invalid or revoked.

During a European opposition proceeding, a request by the patent holder for limitation or revocation is not permitted.[24] If an opposition is filed while a limitation procedure is already underway, the limitation procedure is discontinued and the limitation fee is refunded.[25] An opposition proceeding therefore always takes precedence.

A limitation of the scope of protection can be made using a feature from the description instead of from a dependent claim. Fig. 6.7 visually illustrates such a limitation.

If the main claim is also abandoned, the scope of protection is defined by claims 2, 3, and 4, as shown in Fig. 6.8.

[20] EPO, https://www.epo.org/de/legal/case-law/2022/clr_iii_q_1_1.html, accessed on 19.8.2025.

[21] Article 105a EPC.

[22] Section 64(1) Patent Act or Article 105a(1) sentence 1 EPC.

[23] Section 64(1) Patent Act or EPO, https://www.epo.org/de/legal/guide-epc/2023/ga_c5_6.html, accessed on 21.8.2025.

[24] Article 105a(2) EPC; Rule 93(1) Implementing Regulations to the EPC.

[25] Rule 93(2) sentence 1 Implementing Regulations to the EPC.

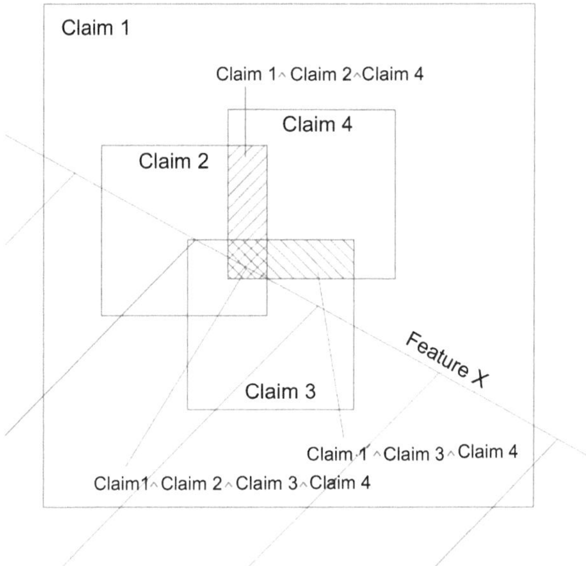

Fig. 6.7 Limitation with a feature from the description

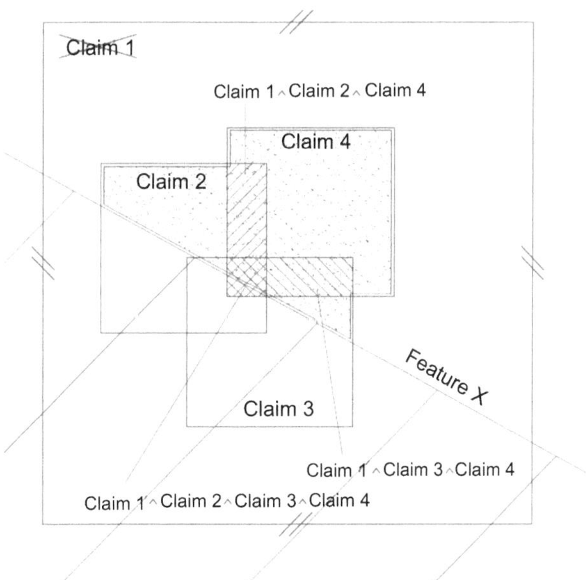

Fig. 6.8 Abandonment of main claim and limitation with a feature

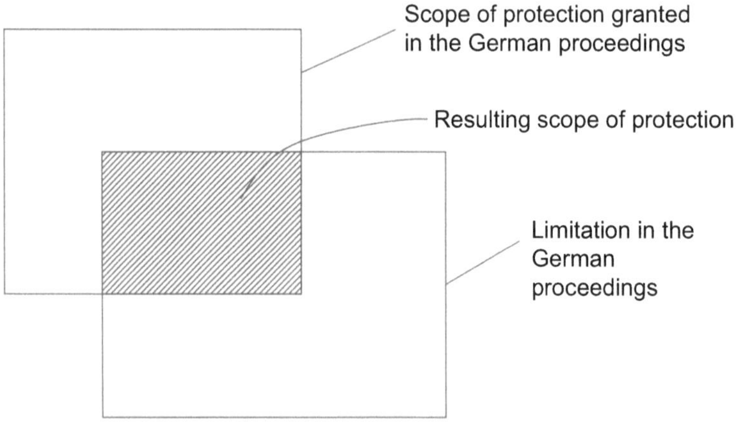

Fig. 6.9 Limitation of a granted patent

6.8 Multiple Limitation of the Patent

Figure 6.9 shows the scope of protection of a set of claims after the limitation of a granted German patent, where only the intersection of the scope of protection after grant and the extent described by the limitation is considered the resulting scope of protection. The rationale for the resulting intersection is that the resulting scope of protection cannot be greater than the granted scope and cannot extend beyond the limitation. If multiple limitations result in no intersection, the patent is not invalidated, but it loses its protective effect.

A European patent can be limited in opposition proceedings and thereafter the European part of the patent can be further limited by the German procedure. The question then arises as to what scope of protection results.

Figure 6.10 shows a multiple limitation, where, for reasons of legal certainty for the public, all limitations are fully effective. The public should not be confronted with a scope of protection that was not claimed in an earlier version. At no time should a previously unclaimed embodiment give rise to a new right to prohibit. An extension of the scope of protection must always be excluded. If there is no common intersection, the patent is left without a resulting scope of protection and is therefore ineffective.

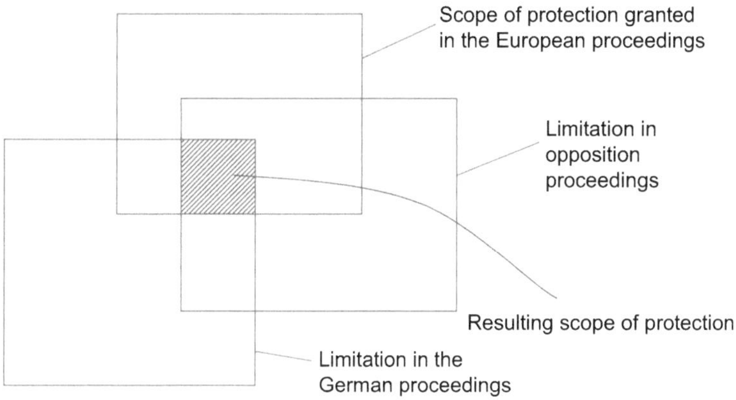

Fig. 6.10 Limitation in the German and European proceedings

6.9 Correction of Errors

The correction of errors is subject to strict limitations under patent law in order to prevent inadmissible extensions and an extension of the scope of protection.

Correction of Errors in German Proceedings
In German proceedings, only formal and clearly recognizable errors can be rectified.[26] Substantive content can never be corrected, even if this would eliminate an ambiguity. The replacement of individual terms is likewise excluded, since even alleged synonyms generally do not embody absolutely identical meanings. Therefore, the use of synonyms cannot rule out an inadmissible extension or an extension of the scope of protection. The correction of errors is generally limited to typographical and calculation errors.

Correction of Errors in European Proceedings
Linguistic errors, typographical errors, and inaccuracies can be corrected in documents filed with the European Patent Office.[27]

The correction of an error in a claim is only possible if the error is immediately and unambiguously apparent to the skilled person from the originally filed application

[26] Section 38 sentence 1 Patent Act.

[27] Rule 139 sentence 1 Implementing Regulations to the EPC.

documents and the claim, and it is obvious how the error is to be corrected.[28] The general knowledge of the skilled person must be taken into account in this context.[29] If general knowledge is required as evidence of the obviousness of the error, it must be substantiated by encyclopedias or standard reference works.[30]

The correction of errors must be of a declaratory nature only and must under no circumstances result in an inadmissible extension or an extension of the scope of protection.[31] The replacement of the entire application documents in the course of a correction procedure is excluded.[32]

6.10 Correction of Translations

At both the German and the European Patent Office, patent applications may be filed in any language.[33] However, translations must be subsequently submitted in each case.

Correction of Translations in German Proceedings

If an application is not filed in German at the German Patent and Trademark Office, the applicant must submit a translation into German within three months of filing the foreign-language application.[34] The translation must be certified by a lawyer or patent attorney with a declaration that the translation is, to the best of their knowledge, correct and complete.[35] Alternatively, the translation may be prepared by a publicly appointed translator.[36]

[28] Rule 139 sentence 2 Implementing Regulations to the EPC; Section 2.2.1, Part H—Chapter VI-3 Guidelines for Examination in the EPO, https://link.epo.org/web/legal/guidelines-epc/de-epc-guidelines-2025-hyperlinked.pdf, accessed on 20.8.2025.

[29] EPO G3/89, 19.11.1992, https://www.epo.org/de/boards-of-appeal/decisions/g890003dx1, accessed on 21.8.2025.

[30] EPO T 881/02, 16.12.2004, https://www.epo.org/en/boards-of-appeal/decisions/t020881eu1, accessed on 21.8.2025.

[31] EPO G 11/91, 19.11.1992, https://www.epo.org/de/boards-of-appeal/decisions/g910011dp1, accessed on 21.8.2025.

[32] EPO G 2/95, 14.5.1996, https://www.epo.org/de/boards-of-appeal/decisions/g950002dx1, accessed on 20.8.2025.

[33] Section 35a paragraph 1 sentence 1 Patent Act and Article 14 paragraph 2 sentence 1 EPC; in contrast, the Austrian Patent Act only permits patent applications in German, English, and French (Section 89 paragraph 2 sentence 2 Austrian Patent Act).

[34] Section 35a paragraph 1 sentence 1 Patent Act.

[35] Section 14 paragraph 1 first alternative Patent Ordinance; Federal Court of Justice, 18.7.2011, X ZB 10/10—Polierendpunktbestimmung, https://juris.bundesgerichtshof.de/cgi-bin/rechtsprechung/document.py?Gericht=bgh&Art=en&sid=64ff5ac465af41817512938ff035c7a7&nr=57672&anz=2&pos=1, accessed on 21.8.2025.

[36] Section 14 paragraph 1 second alternative Patent Ordinance.

The translated application documents must meet the minimum requirements for a patent application to which a filing date can be accorded.[37] In this case, the three-month period for submitting the translated application is deemed to have been met. If a translation of the claims is initially missing, it may also be validly submitted after the three-month period.[38]

Errors in the translation can therefore be corrected at any time. The version that ultimately determines the disclosure content is the originally filed foreign-language application.

Correction of Translations in European Proceedings

A European patent application may be filed in any language, but a translation into one of the official languages—German, English, or French—must be subsequently submitted.[39] A correction of the translation is possible at any time during the proceedings before the European Patent Office, including during opposition proceedings.[40] The correction must bring the translated patent application into conformity with the foreign-language patent application. In opposition proceedings, care must be taken to ensure that the correction does not result in an extension of the scope of protection.[41]

6.11 No Change in the Scope of Protection Despite Additional Features

Features that are subsequently inserted and merely describe already existing features once again do not extend the scope of protection of the patent. Such features therefore do not lead to the inescapable trap and may remain in the relevant claim or be deleted. If the additional feature merely describes the effect of an already existing feature, the same applies. Therefore, no additional technical content may be introduced by subsequently inserted features; only what is already present may be described or explained. In this case, there is neither an inadmissible extension nor an extension of the scope of protection.[42]

[37] Section 35 paragraph 1 Patent Act.

[38] Federal Court of Justice, 18.7.2011, X ZB 10/10—Polierendpunktbestimmung, https://juris.bundesgerichtshof.de/cgi-bin/rechtsprechung/document.py?Gericht=bgh&Art=en&sid=64ff5ac465af41817512938ff035c7a7&nr=57672&anz=2&pos=1, accessed on 21.8.2025.

[39] Article 14 paragraph 2 sentence 1 in conjunction with Article 14 paragraph 1 EPC; in contrast, the Austrian Patent Act only permits patent applications in German, English, and French (Section 89 paragraph 2 sentence 2 Austrian Patent Act).

[40] EPO, https://www.epo.org/de/legal/guidelines-epc/2025/a_vii_7.html, accessed on 20.8.2025.

[41] Article 123 paragraph 3 EPC.

[42] EPO G 3/08, 12.5.2010, OJ EPO 2011, 10; EPO G 1/19. 10.3.2021.

6.12 Change of Category After Grant

A change of category of a patent claim, i.e., the rewording of a method claim into a device claim or the transformation of a device claim into a method claim, after grant is particularly problematic in light of the prohibition on extending the scope of protection.[43]

According to case law, a change of patent category is only permissible if, from the perspective of the skilled person, the new claim formulation is already contained in the previous one and the skilled person would consider it to be encompassed by the old claim.[44]

A change of category of only a dependent claim, while the referenced independent claim remains unaffected, is generally to be rejected, as this creates a new independent claim with a previously non-existent scope of protection. In this case, there is an extension of the scope of protection, which is inadmissible.

[43] EPO 21.10.1987, T 0378/86 (change of category).

[44] EPO, https://www.epo.org/de/legal/guidelines-epc/2025/h_v_7.html, accessed on 10.7.2025.

MIX
Papier aus verantwortungsvollen Quellen
Paper from responsible sources
FSC® C105338

If you have any concerns about our products,
you can contact us on
ProductSafety@springernature.com

In case Publisher is established outside the EU,
the EU authorized representative is:
Springer Nature Customer Service Center GmbH
Europaplatz 3, 69115 Heidelberg, Germany

Printed by Libri Plureos GmbH
in Hamburg, Germany